AF356597

Protection of Future Electricity Systems

Editors

Adam Dyśko
Dimitrios Tzelepis

MDPI • Basel • Beijing • Wuhan • Barcelona • Belgrade • Manchester • Tokyo • Cluj • Tianjin

Editors
Adam Dyśko
Department of Electronic and
Electrical Engineering,
University of Strathclyde
UK

Dimitrios Tzelepis
Department of Electronic and
Electrical Engineering,
University of Strathclyde
UK

Editorial Office
MDPI
St. Alban-Anlage 66
4052 Basel, Switzerland

This is a reprint of articles from the Special Issue published online in the open access journal *Energies* (ISSN 1996-1073) (available at: https://www.mdpi.com/journal/energies/special_issues/ Protection).

For citation purposes, cite each article independently as indicated on the article page online and as indicated below:

LastName, A.A.; LastName, B.B.; LastName, C.C. Article Title. *Journal Name* **Year**, *Volume Number*, Page Range.

ISBN 978-3-0365-3016-1 (Hbk)
ISBN 978-3-0365-3017-8 (PDF)

Cover image courtesy of Adam Dyśko.

Contents

About the Editors

Adam Dyśko (Senior Lecturer) has been with the University of Strathclyde since 1994, first as a PhD student, then as a research fellow, and finally as a member of academic staff. Since 2007, he has been a lecturer in the Department of Electronic and Electrical Engineering, teaching the fundamentals of electrical engineering and power system protection.

He undertakes fundamental research on unconventional power system protection methods for transmission and distribution networks, especially systems with high penetration of renewables. He also provides technical expertise and research support to the industrial working groups and organisations in the UK, such as Grid Code Review Panel (GCRP), Distribution Code Review Panel (DCRP) and Energy Networks Association (ENA).

Dimitrios Tzelepis (Research Fellow) received a BEng (Hons) in Electrical Engineering from Technological Education Institution of Athens (2013), followed by an MSc in Wind Energy Systems (2014), and PhD in Protection, Fault Location and Control in MTDC grids (2017) from the University of Strathclyde. He is currently a Post-Doctoral Researcher with the Department of Electronic and Electrical Engineering, University of Strathclyde. He is a member of IEEE, IET and CIGRE and a regular reviewer for IEEE Journals.

His main research interests lie within the area of power system protection, automation and control of future electricity grids, incorporating increased penetration of renewable energy sources and high voltage direct current interconnections. He is also interested in the application, control and protection of hybrid AC/DC grids including super-conducting feeders, non-homogeneous transmission lines, advanced sensing technologies, and optimised performance of active distribution networks.

Preface to "Protection of Future Electricity Systems"

One of the important areas in powers systems, one which is sometimes overlooked, is power system protection, which is a safety critical component of all electrical systems. Fast and reliable detection of abnormal conditions, as well as identification and prompt removal of the faulty component, help in preserving system integrity, and minimise potential negative impact on the rest of the grid. The consequences of a major failure of the transmission system protection, likely leading to system backout, are difficult even to imagine.

Therefore, there is a pressing need for innovation and a new long-term strategy in power system protection. We hope this Special Issue makes a useful contribution towards this goal. We present ten interesting papers covering a wide range of topics, all related to protection system problems and solutions.

Adaptive protection: An idea of adaptive protection, where relay settings are modified in response to changing system configuration or operating conditions, has been present in technical literature for quite some time. While in the past, its practical application was not perceived as critical, and was additionally limited by technological constraints, the development of numerical multifunctional relays accompanied by digital communications, as well as increased system fault level variability, has enabled adaptive protection to progress to practical applications. In this Special Issue, we include two papers which propose adaptive solutions for both high voltage transmission system and the distribution level microgrids.

Protection of DC systems: There are many situations when DC electricity systems are preferable to AC systems, especially when considering rapid advances in power electronic converter technologies and utilisation of energy sources which naturally produce DC power, such as photovoltaics or batteries. Both HVDC and LVDC systems are being developed and the protection of such networks has a separate set of challenges in terms of ultra-fast fault detection requirements and DC fault current breaking. Therefore, we are pleased to present four papers tackling various aspects of DC network protection.

Enhanced/unconventional protection-systems: There is always a plethora of papers aiming to improve the operation of conventional protective methods, either by utilising advanced processing techniques, or by taking advantage of emerging technological developments. This Special Issue is no exception. You will find some fine examples of unconventional thinking and innovation in most of the presented papers.

Protection of superconducting transmission systems: Due to the introduction of superconducting cables, long-distance superconducting electricity transmission has recently become a reality. Due to their unique physical characteristics, such cables require special attention when it comes to protection. We are pleased to present one paper which specifically addresses this new exciting area of research.

Lightning protection: Last but not least, we have one paper which addresses an important aspect of lightning protection in building integrated photovoltaic modules.

Adam Dyśko, Dimitrios Tzelepis
Editors

"

Editorial

Protection of Future Electricity Systems

Adam Dyśko * and Dimitrios Tzelepis

Department of Electronic and Electrical Engineering, University of Strathclyde, Glasgow G1 1XW, UK; dimitrios.tzelepis@strath.ac.uk
* Correspondence: a.dysko@strath.ac.uk

Citation: Dyśko, A.; Tzelepis, D. Protection of Future Electricity Systems. *Energies* **2022**, *15*, 704. https://doi.org/10.3390/en15030704

Received: 14 December 2021
Accepted: 17 January 2022
Published: 19 January 2022

Publisher's Note: MDPI stays neutral with regard to jurisdictional claims in published maps and institutional affiliations.

1. Introduction

The electrical energy industry is undergoing dramatic changes; the massive deployment of renewables, an increasing share of DC networks at transmission and distribution levels, and at the same time, a continuing reduction in conventional synchronous generation, all contribute to a situation where a variety of technical and economic challenges emerge. As society's reliance on electrical power continues to increase as a result of international decarbonisation commitments, the need for the secure and uninterrupted delivery of electrical energy to all customers has never been greater.

One of the important areas, sometimes overlooked, is power system protection, which is a safety-critical component of all electrical systems. The fast and reliable detection of abnormal conditions, as well as the identification and prompt removal of faulty components, help in preserving system integrity and minimise the potential negative impact on the rest of the grid. The consequences of a major failure of the transmission system protection, likely leading to system backout, are difficult even to imagine.

Therefore, there is a pressing need for innovation and a new, long-term strategy in power system protection. We hope this Special Issue makes a useful contribution towards this goal. We present 10 interesting papers that cover a wide range of topics, all related to protection system problems and solutions.

2. Short Review of Contributions

Adaptive protection: The idea of adaptive protection, where relay settings are modified in response to changing system configuration or operating conditions, has been present in the technical literature for quite some time. While in the past its practical application was not perceived as critical and was also limited by technological constraints, the development of numerical multifunctional relays accompanied by digital communications, as well as increased system-fault-level variability, has enabled adaptive protection to progress to practical applications. In this issue, we include two papers [1,2] which propose adaptive solutions for both high-voltage transmission systems and distribution-level microgrids.

Protection of DC systems: There are many situations where DC electricity systems are preferable to AC systems, especially when considering rapid advances in power electronic converter technologies and the utilisation of energy sources which naturally produce DC power, such as photovoltaics or batteries. Both HVDC and LVDC systems are being developed, and the protection of such networks has a separate set of challenges in terms of ultra-fast fault detection and classification requirements, fault-ride through capabilities, and DC fault current breaking. Therefore, we are pleased to present four papers [3–6] that tackle various aspects of DC network protection.

Enhanced/unconventional protection-systems: There is always a plethora of papers that aim to improve the operation of conventional protective methods, either by utilising advanced processing techniques or by taking advantage of emerging technological developments (e.g., distributed sensing technologies). This issue is no exception. You will find some fine examples of unconventional thinking and innovation in most of the presented papers (e.g., [3,5,7,8], to highlight a few).

Protection of superconducting transmission systems: Due to the introduction of superconducting cables, long-distance, superconducting electricity transmission has recently become a reality. Due to their unique physical characteristics, such cables require special attention when it comes to protection. We are pleased to present one paper [9] which specifically addresses this new, exciting area of research.

Lightning protection: Last but not least, we present one paper which addresses an important aspect of lightning protection in building integrated photovoltaic modules [10].

3. Conclusions

We sincerely hope the papers included in this Special Issue will inspire both academics and protection system practitioners to further develop much-needed solutions for the safe operation of future electricity systems. We strongly believe that there is a need for more work to be carried out, and we hope this issue provides a useful open-access platform for the dissemination of new ideas, as well as a catalyst for further protection innovation.

Author Contributions: Conceptualization, A.D. and D.T.; writing—original draft preparation, A.D.; writing—review and editing, A.D. and D.T.; All authors have read and agreed to the published version of the manuscript.

Funding: This research received no external funding.

Conflicts of Interest: The authors declare no conflict of interest.

References

1. Baayeh, A.G.; Bayati, N. Adaptive Overhead Transmission Lines Auto-Reclosing Based on Hilbert–Huang Transform. *Energies* **2020**, *13*, 5416. [CrossRef]
2. Memon, A.A.; Kauhaniemi, K. An Adaptive Protection for Radial AC Microgrid Using IEC 61850 Communication Standard: Algorithm Proposal Using Offline Simulations. *Energies* **2020**, *13*, 5316. [CrossRef]
3. Yu, J.; Zhang, Z.; Xu, Z. A Local Protection and Local Action Strategy of DC Grid Fault Protection. *Energies* **2020**, *13*, 4795. [CrossRef]
4. Kim, G.; Lee, J.S.; Park, J.H.; Choi, H.D.; Lee, M.J. A Zero Crossing Hybrid Bidirectional DC Circuit Breaker for HVDC Transmission Systems. *Energies* **2021**, *14*, 1349. [CrossRef]
5. van der Blij, N.H.; Purgat, P.; Soeiro, T.B.; Ramirez-Elizondo, L.M.; Spaan, M.T.J.; Bauer, P. Decentralized Plug-and-Play Protection Scheme for Low Voltage DC Grids. *Energies* **2020**, *13*, 3167. [CrossRef]
6. Huang, W.; Luo, G.; Cheng, M.; He, J.; Liu, Z.; Zhao, Y. Protection Method Based on Wavelet Entropy for MMC-HVDC Overhead Transmission Lines. *Energies* **2021**, *14*, 678. [CrossRef]
7. Ji, L.; Cao, Z.; Hong, Q.; Chang, X.; Fu, Y.; Shi, J.; Mi, Y.; Li, Z. An Improved Inverse-Time Over-Current Protection Method for a Microgrid with Optimized Acceleration and Coordination. *Energies* **2020**, *13*, 5726. [CrossRef]
8. Yin, X.; Wang, Y.; Qiao, J.; Xu, W.; Yin, X.; Jiang, L.; Xi, W. Multi-Information Fusion-Based Hierarchical Power Generation-Side Protection System. *Energies* **2021**, *14*, 327. [CrossRef]
9. Tsotsopoulou, E.; Dyśko, A.; Hong, Q.; Elwakeel, A.; Elshiekh, M.; Yuan, W.; Booth, C.; Tzelepis, D. Modelling and Fault Current Characterization of Superconducting Cable with High Temperature Superconducting Windings and Copper Stabilizer Layer. *Energies* **2020**, *13*, 6646. [CrossRef]
10. Bian, X.; Zhang, Y.; Zhou, Q.; Cao, T.; Wei, B. Numerical and Experimental Study of Lightning Stroke to BIPV Modules. *Energies* **2021**, *14*, 748. [CrossRef]

Article

Decentralized Plug-and-Play Protection Scheme for Low Voltage DC Grids

Nils H. van der Blij [1,*], **Pavel Purgat** [1], **Thiago B. Soeiro** [1], **Laura M. Ramirez-Elizondo** [1], **Matthijs T. J. Spaan** [2] **and Pavol Bauer** [1]

[1] Department of Electrical Sustainable Energy, Delft University of Technology, Mekelweg 4, 2628 CD Delft, The Netherlands; P.Purgat@tudelft.nl (P.P.); T.BatistaSoeiro@tudelft.nl (T.B.S.); L.M.RamirezElizondo@tudelft.nl (L.M.R.-E.); P.Bauer@tudelft.nl (P.B.)

[2] Department of Software Technology, Delft University of Technology, Van Mourik Broekmanweg 6, 2628 XE Delft, The Netherlands; M.T.J.Spaan@tudelft.nl

* Correspondence: N.H.vanderBlij@TUDelft.nl

Received: 22 April 2020; Accepted: 9 June 2020; Published: 18 June 2020

Abstract: Since the voltages and currents in dc grids do not have a natural zero-crossing, the protection of these grids is more challenging than the protection of conventional ac grids. Literature presents several unit and non-unit protection schemes that rely on communication, or knowledge about the system's topology and parameters in order to achieve selective protection in these grids. However, communication complicates fast fault detection and interruption, and a system's parameters are subject to uncertainty and change. This paper demonstrates that, in low voltage dc grids, faults propagate fast through the grid and interrupted inductive currents commutate to non-faulted sections of the grid, which both can cause circuit breakers in non-faulted sections to trip. A decentralized plug-and-play protection scheme is proposed that ensures selectivity via an augmented solid-state circuit breaker topology and by utilizing the proposed time-current characteristic. It is experimentally shown that the proposed scheme provides secure and selective fault interruption for radial and meshed low voltage dc grids under various conditions.

Keywords: decentralized protection scheme; fault analysis; low voltage direct current grids; plug-and-play systems; solid-state circuit breakers

---

1. Introduction

Low voltage dc grids have gained attention due to the potential advantages over low voltage ac grids when power electronic devices are proliferated in the system. Firstly, in such systems, the required number of conversion steps is generally reduced leading to improved system efficiency. Secondly, because the switching frequencies of power electronic converters are typically much higher than the native 50/60 Hz frequency of ac grids, the size of passive components can be reduced. Lastly, the absence of frequency and phase can make the control of dc grids significantly simpler [1–5].

The protection of low voltage dc grids is more challenging than the protection of conventional low voltage ac systems. Fundamentally, it is more difficult to interrupt inductive currents and extinguish arcs, since the voltages and currents in dc grids do not have a natural zero crossing [6,7]. Furthermore, these grids are often meshed and subjected to bi-directional power flows, complicating the detection and selectivity compared to conventional radial networks [8]. Moreover, in order to prevent high fault currents and blackouts, low voltage dc grids usually require fast fault interruption [9,10].

Because of the limited overload capability of power electronic devices, being able to withstand short-circuit conditions for milliseconds leads to oversized components in terms of current-carrying capability [10–12]. Furthermore, for dc systems with low inertia, a blackout is inevitable when a fault is sustained for a longer period of time. Fuses, electromechanical devices and hybrid circuit breakers provide solutions for clearing faults in the order of milliseconds to seconds, but faster fault detection and interruption is required for low voltage dc systems [10,13–15]. To achieve this, the low voltage dc systems can be protected with solid-state circuit breakers (SSCBs), which can detect and interrupt faults within microseconds [16,17].

Several non-unit and unit protection schemes for low voltage dc grids have been reported in literature [18–20]. Non-unit protection schemes utilize local measurements in order to detect faults. Many of these protection schemes measure the current and current rate-of-change, and circuit breakers are opened when preset thresholds are exceeded, but the utilization of higher order derivatives of the current and the grid's voltage are also reported [21–23]. The main advantages of non-unit protection schemes are their simplicity, and their resilience to the failure of protection devices when a hierarchical structure of circuit breakers is used. However, these schemes have difficulty isolating only the faulted areas of the grid and thus achieving selectivity. Therefore, protection schemes were proposed that utilize knowledge about the system's topology in order to achieve selectivity. For example, faults can be located by measuring the grid's impedance and comparing it to known line parameters, or a wavelet transform can be used to identify faults by comparing them to simulations of the system [24–28]. Furthermore, a handshaking protection scheme was introduced, which locates and isolates a fault by temporarily powering down the dc system [29]. Nevertheless, these methods struggle to ensure selectivity when system parameters are uncertain or the system topology is changing. On the other hand, unit protection schemes achieve selectivity by utilizing a communication infrastructure. For instance, differential protection schemes locate faults by comparing the currents at different locations in the system, and event-based protection schemes ensure selectivity by combining local detection with central decision-making [30–36] However, since fast fault detection and interruption is required in low voltage dc grids, utilizing a communication infrastructure is not desirable.

The main contribution of this paper is a decentralized plug-and-play protection scheme, which contrary to other protection schemes, ensures selectivity without utilizing communication and only requires minimal knowledge about the system. Selectivity is achieved by augmenting the standard solid-state circuit breaker topology with RC dampers and by utilizing the proposed time-current characteristic. The protection scheme is plug-and-play in the sense that selective protection is provided on both sides of the circuit breakers in the system, regardless of the system's configuration or where the circuit breakers are located in the system and without requiring (re)configuration of the circuit breakers. Furthermore, the protection scheme is experimentally validated, showing the effectiveness of the protection scheme for different low voltage dc systems under various conditions.

This paper is structured as follows. In Section 2, it is discussed that current limiting inductances and fast fault interruption are crucial for the protection of low voltage dc grids. In Section 3, the experimental setup is presented and the operation of the designed solid-state circuit breaker (SSCB) is validated. In Section 4, it is experimentally shown that fast fault propagation and the commutation of inductive currents pose two challenges for the selectivity of non-unit protection schemes. In Section 5, it is proposed to add an RC damper to the output terminals of the SSCBs in order to delay fault propagation and smoothe current commutation. Furthermore, a time-current characteristic is proposed that coordinates upstream and downstream circuit breakers and also prevents tripping due to current commutation. In Section 6, the proposed plug-and-play protection scheme is experimentally validated. Finally, in Section 7, conclusions are drawn.

2. Short-Circuit Fault Currents in Low Voltage DC Grids

In low voltage dc grids overvoltages can occur when, for instance, lightning strikes one of the conductors. Therefore, surge arresters such as Metal Oxide Varistors (MOVs) or spark gaps should be used to clamp the voltage. Furthermore, short-circuits can occur when, for example, a tree falls on one of the overhead lines or the insulation deteriorates in one of the underground lines. In those cases, one or more conductors are short-circuited to each other or to the ground [37].

To calculate the short-circuit fault current in a monopolar dc grid, the equivalent circuit in Figure 1 is used [18,21,25]. The fault current is highest when the voltage on the non-faulted part of the system remains constant, and therefore this part of the system is modelled by a voltage source U_{dc}. Furthermore, the SSCB is modelled by an ideal switch, its on-state resistance R_{CB} and its (intrinsic) inductance L_{CB}. Since the lines in low voltage dc grids are short, the propagation delay can be neglected and lumped element models are sufficiently accurate [38]. Therefore, the overhead or underground line(s) between the SSCB and the short-circuit are modelled by a lumped element π-model.

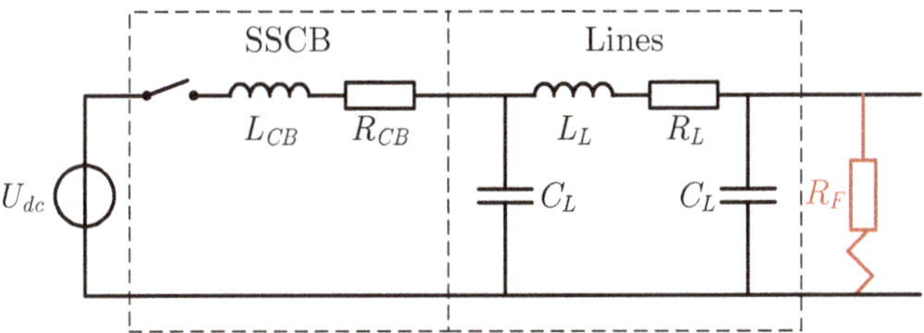

Figure 1. Equivalent circuit to calculate the worst-case short-circuit fault current in dc grids.

Simulation results for the current during a low resistance fault (0.1 Ω) and a high resistance fault (10 Ω) are shown in Figure 2. The fault current is shown for different lengths of the distribution line between the SSCB and the fault, which have a typical resistance of 1 Ω/km, an inductance of 0.25 mH/km and a capacitance of 0.5 μF/km. Furthermore, during these simulations the grid voltage U_{dc} is 350 V, the on-resistance R_{CB} is 0.1 Ω, and the SSCB's inductance L_{CB} is 1 μH.

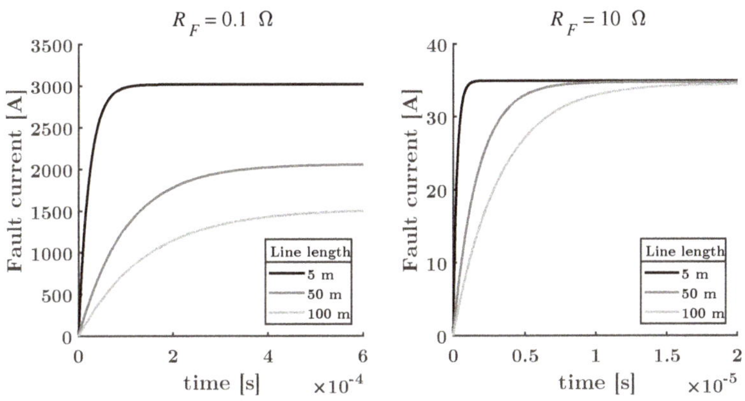

Figure 2. Simulation results for the fault current in the equivalent circuit of Figure 1 for different fault resistances and distribution line lengths.

Since the capacitance of the line C_L is small, the fault current can be approximated by

$$I_F(t) = \frac{U_{dc}}{R_{CB} + R_L + R_F} \left(1 - e^{-\frac{R_{CB}+R_L+R_F}{L_{CB}+L_L}t}\right),\tag{1}$$

where R_F is the resistance of the fault.

Note that the steady-state fault current is only determined by the total resistance, which is the reason short-circuit currents are so high in dc grids. Moreover, the line length only has a significant influence on the steady-state current when the fault resistance is low. Furthermore, by differentiating (1) it becomes clear that the current rate of change is only determined by the sum of the inductances in the system.

The thermal and electrical design of the SSCBs and other components in the grid are dependent on the duration and magnitude of the worst-case fault current that they need to be able to sustain. In the worst case, the short-circuit occurs close to the terminals of the SSCB, making the total inductance close to L_{CB}. Furthermore, SSCB's are designed to have as low on-state resistance as possible in order to improve the system's efficiency. Therefore, if the current before the fault was the nominal current I_{nom}, the worst-case fault current can be approximated by

$$I_{F,\max} = \frac{U_{dc}t_{\max}}{L_{CB}} + I_{nom},\tag{2}$$

where $t_{\max}$ is the maximum time that the SSCB needs to detect the fault and open its switches.

From (2) it is clear that, in order to reduce the worst-case fault current, fast fault detection and interruption are essential. Furthermore, even though SSCBs can detect and clear faults within 1 µs, a current limiting inductance is often added to SSCBs in order to further limit the maximum fault current. For example, assuming a grid voltage of 350 V, an SSCB clearing time of 1 µs, a nominal current of 20 A, and a current limiting inductance of 1 µH, the maximum fault current is 370 A.

Since the worst-case fault current develops when the short-circuit occurs at the SSCB's terminals, this worst-case fault current is not dependent on the system's parameters or uncertainty in the system. Furthermore, pole-to-pole faults in (grounded) unipolar and bipolar grids exhibit similar behavior to the behavior described in this section, although the resistance and inductance of the return path has to be taken into account. However, because ground faults in these grids have an identical equivalent circuit and behavior, the maximum fault currents in these grids are the same.

3. Experimental Setup

In this paper, the non-unit protection of different low voltage dc grids with multiple power electronic converters, distribution lines and SSCBs is investigated. In this section, the specifications of the different components that make up the experimental setup are presented. Moreover, the operation of the developed SSCBs is experimentally validated.

3.1. Power Electronic Converters Emulated by DC Power Supplies

The power electronic converters in low voltage dc grids are emulated by SM 500-CP-90 power supplies, which are manufactured by Delta Elektronika. The SM 500-CP-90 power supply is an isolated bidirectional power supply rated for 500 V and 90 A in two quadrants. To emulate the behavior of power electronic converters in low voltage dc grids, these power supplies are operated in constant voltage or constant current mode. In these modes, the power supplies internally limit the current, but they do not limit the current flowing from their output capacitances. This accurately emulates power electronic converters, since their control bandwidth is generally low and the circuit breakers detect and interrupt the fault current before the converters' control can react to the disturbance. A picture of such a power supply is shown in Figure 3.

Figure 3. SM 500-CP-90 bidirectional power supply from Delta Elektronika, which is used to emulate the power electronic converters in low voltage dc grids.

3.2. Data Acquisition

The measurements in the experimental setup are acquired by a Yokogawa DLM2034 scope operating at a sample rate of 250 MHz. Furthermore, the voltages are measured by utilizing a Yokogawa 700924 differential voltage probe, which is able to measure up to 1400 V with a bandwidth of 100 MHz. Additionally, the currents are measured by utilizing a Yokogawa 701929 current probe, which can measure currents up to 30 A with a bandwidth of 50 MHz. A picture of these components is shown in Figure 4.

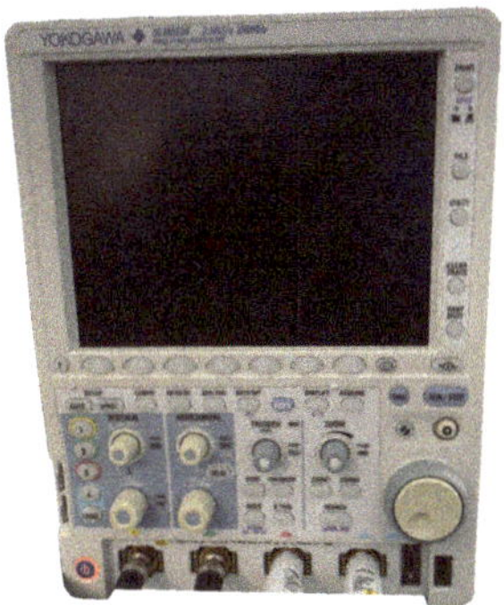

Figure 4. Yokogawa scope, voltage probe and current probe that are used for the data acquisition.

3.3. Distribution Line Emulation Circuit

Due to practical considerations, distribution lines are emulated by a π-equivalent circuit, which is shown in Figure 5. The equivalent circuit emulates a 100 m line with an inductance L_l of 32 µH, a resistance R_l of 120 mΩ and a capacitance C_l of 42 nF. In general, such lumped element models are valid when the wavelength of the signals in the system are longer than the length of the line [38]. Since the length of the lines in this paper are relatively short, this is assumed to be valid. However, this assumption might not be valid for long transmission lines in medium or high voltage systems.

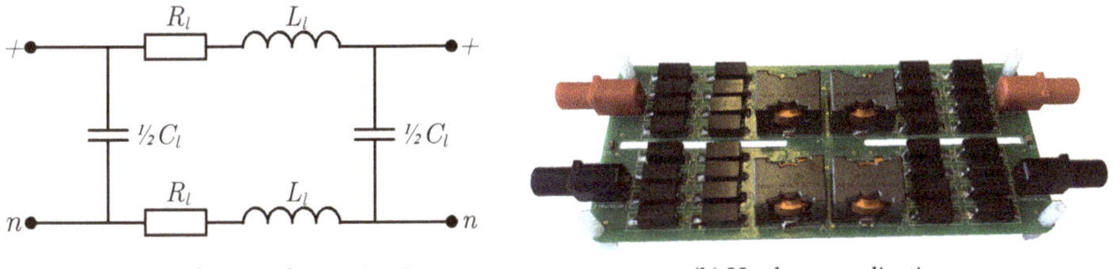

(**a**) π-equivalent emulation circuit   (**b**) Hardware realization

Figure 5. Lumped element emulation circuit that is used to emulate a 100 m distribution line.

3.4. Solid-State Circuit Breaker

The base design of the SSCBs that were developed to investigate non-unit protection schemes is shown in Figure 6. To interrupt the various short-circuit faults, two anti-series SiC (Cree C3M0065090D) switches are used for both the positive pole and the neutral. Furthermore, to prevent an avalanche breakdown of the switches and overvoltages in the grid, Metal Oxide Varistors (MOVs) are used to clamp the voltage. The design parameters of the SSCB are given in Table 1.

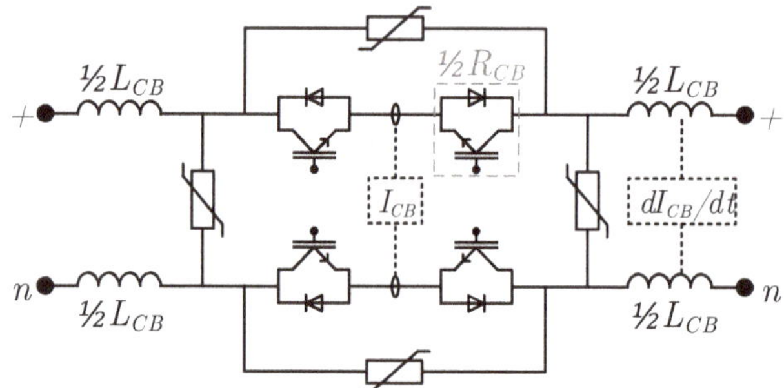

Figure 6. Base design of the solid-state circuit breakers that are used in this paper.

Table 1. Design parameters of the solid-state circuit breaker.

Parameter	Acronym	Value
Nominal voltage	U_{nom}	350 V
Nominal current	I_{nom}	10 A
On-state resistance per pole	R_{CB}	130 mΩ
Current limiting inductance	L_{CB}	1.0 μH
Maximum clearing time	t_{max}	1.0 μs

The SSCB measures the current via a high bandwidth hall-sensor, and the current rate-of-change (di/dt) via the voltage across the current limiting inductor. Using analog comparators, logical gates, and a latch circuit, the switches are turned off when the current through the SSCB or the voltage across the inductor exceed their set thresholds. It will be shown that it is able to detect and open its switches within 1 μs after its thresholds are exceeded. For illustrative purposes, a picture of the SSCB's hardware realization is shown in Figure 7.

Figure 7. Hardware realization of the solid-state circuit breaker.

3.5. Experimental Validation of the SSCB

To validate the operation of the developed SSCB, one side is connected to a voltage source of 350 V while a short-circuit is induced at the other side using a mechanical relay and a variable resistor, much like the circuit shown in Figure 1. For the experiments, the thresholds for the overcurrent and inductor voltage (di/dt) detection are set to 21 A and 20 V (20 MA/s) respectively. For these experiments thresholds can be chosen arbirarily, but guidelines for determining appropriate thresholds will be given in Section 5.

To show the correct operation of the overcurrent detection, the SSCB is short-circuited at its terminal with a relatively high fault resistance and low inductance (8 Ω and 0 µH respectively). The fault current I_F and the voltage over the current limiting inductor U_L for this experiment are shown in Figure 8. At the fault occurrence the di/dt is high, but because the analog detection circuits use small filter capacitors and the system's time constant is low (due to the large fault resistance), the voltage over the inductor does not exceed its 20 V threshold long enough to trip the di/dt detection circuit. However, when the fault current exceeds the 21 A threshold, overcurrent is detected by the analog control logics and the switches are opened within 1 µs.

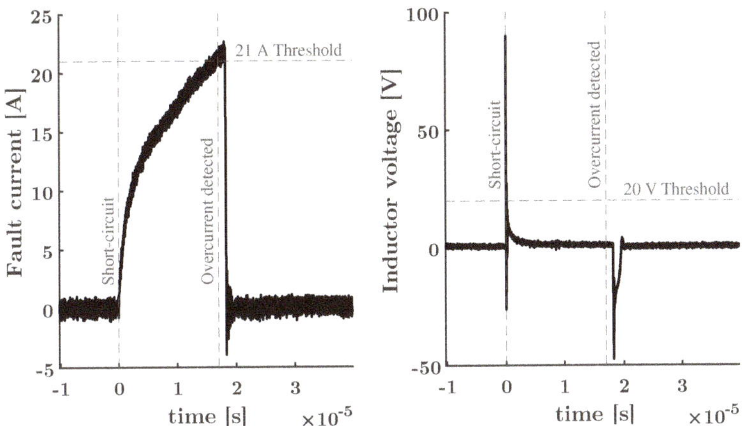

Figure 8. Experimental results when the SSCB is short-circuited with a high fault resistance resulting in the overcurrent detection being triggered when the current exceeds 21 A.

To show the adequacy of the di/dt detection, the experiment is repeated with relatively low fault resistance (2 Ω). The results for this experiment are shown in Figure 9. Because the system's time constant is lower, the voltage over the inductor remains above the threshold significantly longer. Therefore, the analog di/dt detection is triggered and the fault is cleared within 400 ns of its occurrence.

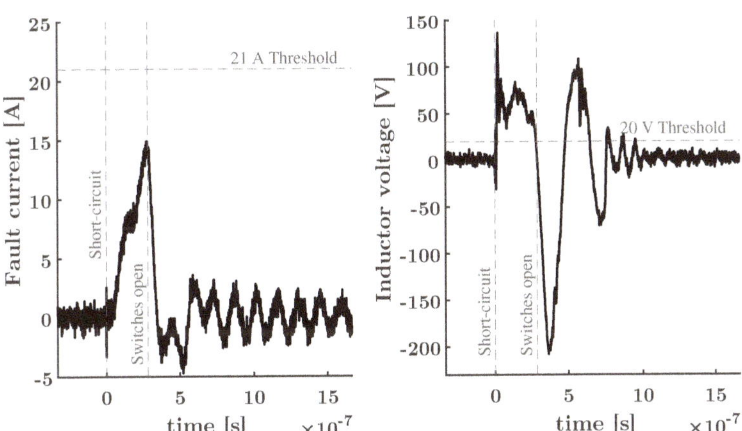

Figure 9. Experimental results when the SSCB is short-circuited with a low fault resistance resulting in the di/dt detection being triggered when the 20 V (20 MA/s) threshold is exceeded for a longer time.

From these two experiments it can be concluded that both the overcurrent and di/dt detection circuits operate adequately, and the SSCB clears faults within 1 μs. In the remainder of this paper three of these SSCBs will be used to experimentally validate the theory presented in this paper.

4. Non-Unit Protection Scheme Challenges

In the previous section it was shown that faults can be cleared within 1 μs by utilizing overcurrent and current rate-of-change detection in combination with an SSCB. Fast and robust fault interruption is possible with such an approach, since no communication infrastructure is utilized. However, in this section it will be shown that achieving selectivity can be challenging when using these non-unit protection methods.

4.1. Fast Fault Propagation

Although the SSCBs current limiting inductance ensures a maximum fault current magnitude, it does not always prevent the fault from propagating through the system and tripping multiple protection devices. To show this the experimental setup shown in Figure 10 is used.

In this setup, a constant voltage source of 350 V and two constant current loads of 5 A are connected to a low inductive dc bus via three SSCBs. This situation can occur, for example, in a dc household that is disconnected from the main grid, where the photovoltaic (PV) panels are providing the energy for loads in two other groups inside the house.

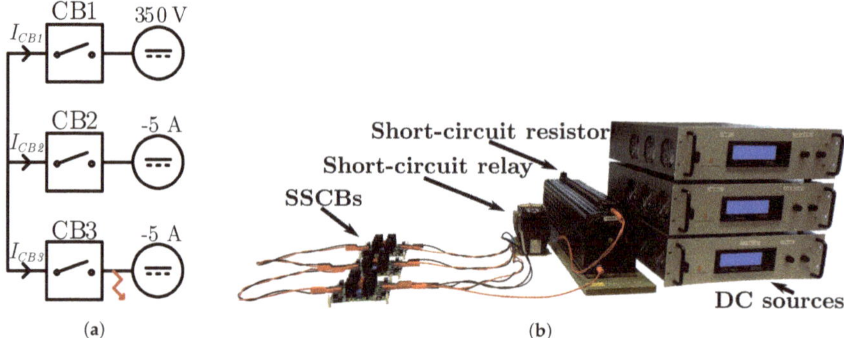

Figure 10. (**a**) Schematic and (**b**) picture of the experimental setup connecting a constant voltage source to two constant current loads through three SSCBs.

To show that, in some cases, the fault propagates through the system and trips all the SSCBs before the SSCB in the faulted group can react, a short-circuit with a very low fault resistance (0.75 Ω) is induced at the load-side terminal of CB3. The experimental results for the voltage over the current limiting inductance of CB2 U_{L2} and the currents flowing in each circuit breaker are shown in Figure 11.

Observe that after the short-circuit is induced, the fault current starts flowing from the converters' output capacitances to the fault. Therefore, the current in CB3 is increasing rapidly, while the currents in CB2 and CB1 are decreasing rapidly. Also note that, although CB2 feeds a load, the discharge of the load converter's capacitance contributes to the fault current. Furthermore, even though the fault occurs at the load side of CB3, the voltage over the current limiting inductance and CB2 exceeds its threshold before CB3 can act and selectivity is lost.

It is important to realize this is not a consequence of utilizing di/dt detection. If only overcurrent detection is used, the currents in CB1 and CB2 would exceed their limits by the time CB3 clears the fault, because of the high current rate-of-change. Therefore, a challenge for the selectivity of non-unit protection schemes is the fast propagation of low impedance faults through low inductive sections of grids. In radial grids, directional detection can be used to overcome this challenge, but for meshed grids this does not work.

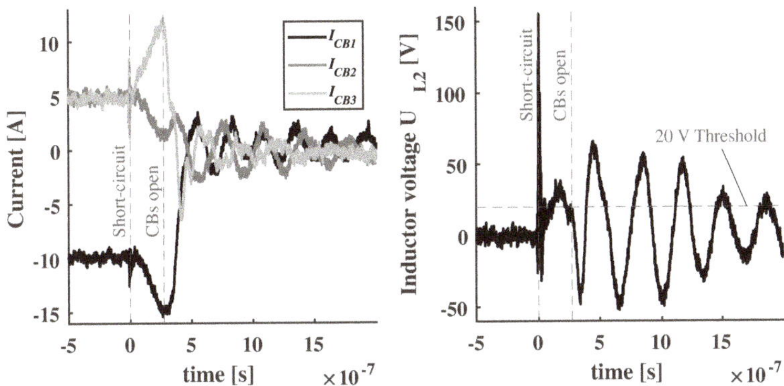

Figure 11. Experimental results for the system shown in Figure 10, showing that fault propagation can cause unnecessary tripping in low inductive systems.

4.2. Commutation of Inductive Currents

If the current has an alternative path, the interrupting SSCB does not always dissipate the excess inductive energy after clearing the fault. Consequently, for a transient period, the inductive current will flow through the remainder of the system, which can trip other SSCBs in the system. To show this the experimental setup shown in Figure 12 is used.

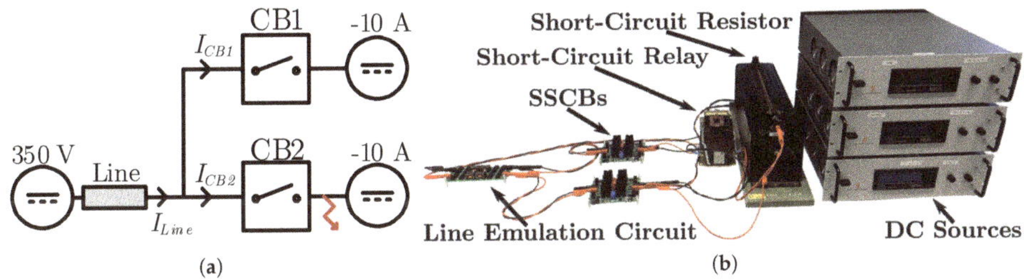

Figure 12. (a) Schematic and (b) picture of the experimental setup connecting a constant voltage source and two constant current loads connected through an inductive line.

For this experiment, a constant voltage source of 350 V is connected to two constant current loads, each consuming 10 A, via an inductive line and two SSCBs. This situation can occur, for example, when a dc household is connected to a main grid. The line in this experiment is emulated by the π-equivalent emulation circuit presented in Section 3.

To show that, in some cases, commutated inductive currents can trip SSCBs in non-faulted parts of the system, a short-circuit with a short-circuit resistance of 4.0 Ω is induced at the load side of CB2. The current in the line and the currents in the circuit breakers for this experiment are shown in Figure 13.

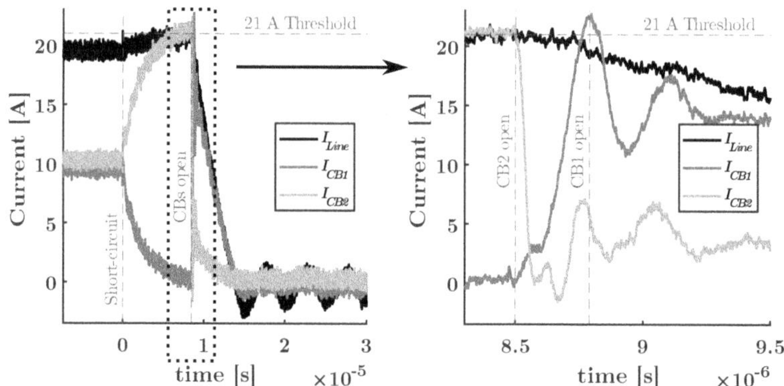

Figure 13. Experimental results for the system shown in Figure 12, showing that the commutation of inductive currents can cause unnecessary tripping (the right figure is a zoom in).

Note that, after the short-circuit occurs, the fault current starts flowing from the output capacitances of the converters. Therefore, the currents in the line and CB2 rise quickly, while the current in CB1 decreases rapidly, until the threshold of CB2 is exceeded. Subsequently, after CB2 opens, the inductive current in the line (that was first shared by CB1 and CB2) is commutated to CB1 almost immediately and its di/dt detection is tripped and selectivity is lost. However, if the di/dt measurement was not tripped, the overcurrent detection would have also been tripped since the current through CB1 also briefly exceeds 21 A. Afterwards, since the inductive line current does not have an alternative path, the inductive energy is dissipated in CB1's MOV's.

When SSCBs are operating near their rated current, the commutation of inductive currents would likely cause a cascade tripping circuit breakers in the system. Moreover, this challenge cannot be solved by directional detection, even in radial systems. Therefore, the commutation of inductive currents poses a challenge for the selectivity of non-unit protection schemes.

5. Proposed Plug-and-Play Protection Scheme

The previous section presented two challenges for the selectivity of non-unit protection schemes. These challenges can be tackled by utilizing communication, but communication will likely slow down fault detection. Furthermore, (directional) thresholds could be designed to prevent unnecessary tripping, but doing so would require knowledge about the system's topology and parameters. Therefore, in order to achieve selective protection for plug-and-play low voltage dc grids, an alternative approach is proposed.

5.1. Proposed SSCB Topology to Delay Fault Propagation

It is proposed to append the SSCB topology with an RC damper on each terminal, as is shown in Figure 14. The purpose of the dampers' capacitance is to temporarily provide a low impedance path for fault currents and commutated inductive currents, delaying their propagation. Consequently, before a fault current can flow on the non-faulted side of the SSCB the damper capacitor on the non-faulted side of the SSCB must be discharged through the current limiting inductance. Similarly, a (commutated) current must first charge the damper capacitor before current can flow in the current limiting inductance. Therefore, the propagation of these currents is delayed, providing time for the SSCBs in the faulted areas to clear the fault and smoothing the commutation of inductive currents. However, if just a capacitance was added, high frequency oscillations with low damping could occur between the damper capacitors through the current limiting inductance, since the on-state resistance of the switches is small. Therefore, resistances are added to the dampers in order to attenuate these oscillations.

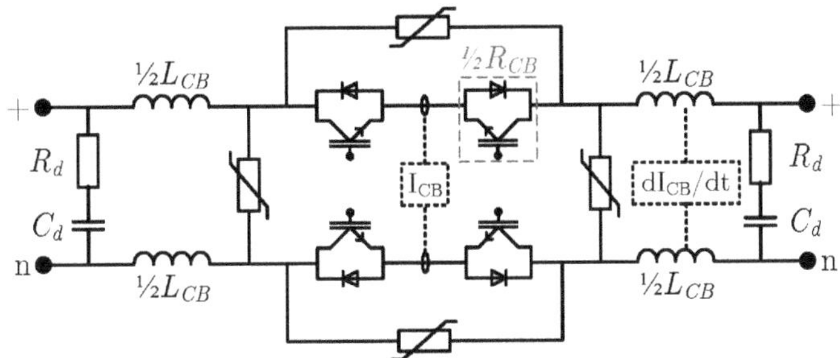

Figure 14. Proposed solid-state circuit breaker topology with added RC dampers.

In the proposed topology, the RC dampers together with the current limiting inductance essentially form a low-pass LCR filter. The inductance, resistance and capacitance of this LCR filter are $2L_{CB}$, $2R_d + 2R_{CB}$ and $\frac{1}{2}C_d$ respectively. Making a loop inside the SSCB, the sum of the voltages over the damper capacitors, damper resistors, on-state resistances and current limiting inductances must be zero. Therefore, the differential equation for the inductor current is given by

$$2L_{CB}\frac{\partial}{\partial t}I(t) + (2R_d + 2R_{CB})I(t) + \frac{2}{C_d}\int I(t)dt = 0. \tag{3}$$

Differentiating this equation, and dividing by $2L_{CB}$ yields

$$\frac{\partial^2}{\partial t^2}I(t) + \frac{R_d + R_{CB}}{L_{CB}}\frac{\partial}{\partial t}I(t) + \frac{1}{L_{CB}C_d}I(t) = 0. \tag{4}$$

Applying the Laplace transform on this equation results in

$$s^2 I(s) + \frac{R_d + R_{CB}}{L_{CB}}sI(s) + \frac{1}{L_{CB}C_d}I(s) = I'(0), \tag{5}$$

where s is a complex variable representing attenuation and frequency in the Laplace domain ($s = \sigma + j\omega$), and $I'(0)$ is the initial condition for the first derivative of the current. Consequently, the transfer function of this system is given by

$$H(s) = \frac{I(s)}{I'(0)} = \frac{1}{s^2 + \frac{R_d + R_{CB}}{L_{CB}}s + \frac{1}{L_{CB}C_d}}. \tag{6}$$

The resonant frequency f_r and attenuation frequency α of this standard second-order system are

$$f_r = \frac{1}{2\pi\sqrt{L_{CB}C_d}}, \tag{7}$$

$$\alpha = \frac{R_d + R_{CB}}{4\pi L_{CB}}, \tag{8}$$

which will be used later in this section to provide design guidelines for the damper parameters.

Note that, the higher damper capacitor, the lower the resonant frequency of the SSCB's LCR circuit and the longer the SSCB will delay the propagation of fault currents. From a different perspective, a higher damper capacitance can provide the energy for the fault current for a longer time.

An additional benefit of the damper capacitance is that it delays and smoothes the commutation of an (inductive) current. This is illustrated by simulating the inductor current in the circuit from Figure 15. For the simulations the grid voltage U_{dc} is 350 V, the on-resistance R_{CB} is 0.1 Ω, the SSCB's inductance L_{CB} is 1 µH, and the damper resistance R_d is 2 Ω. The simulation results for the current in the SSCB's inductors, when the current I_o is stepped up from 0 to 10 A at $t = 0$, are given in Figure 16.

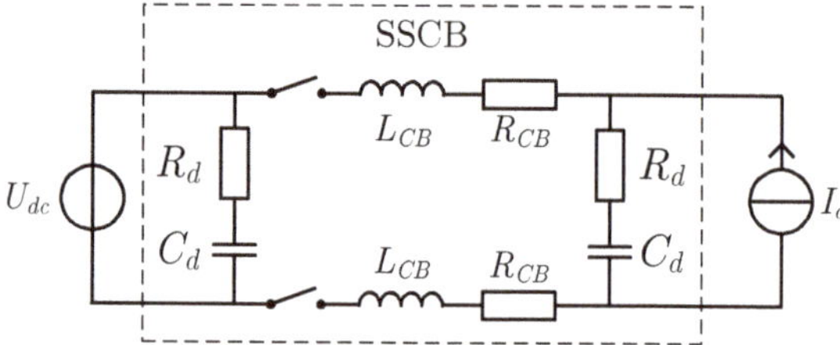

Figure 15. Circuit that is used to show the effect of the RC dampers on the commutation of an (inductive) current.

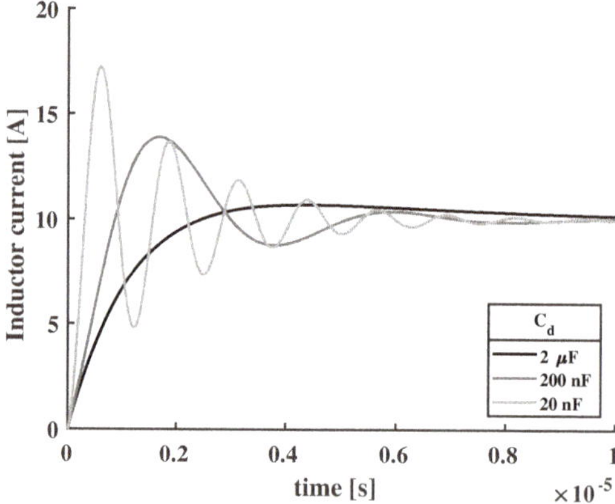

Figure 16. Simulation results for the inductor current in the circuit of Figure 15 for different damper capacitances.

The simulation results show that the damper capacitance absorbs the forced current, delaying the current from flowing inside the SSCB. It also illustrates that, at lower damper capacitances, current overshoot and underdamped high frequency oscillations can occur.

5.2. Proposed Time-Current Characteristic

In ac systems selective coordination between upstream and downstream circuit breakers is often achieved in radial systems by using time-current characteristics for the protection devices. Time-current characteristics depict how long a protection device allows a current to flow before it interrupts it, and they are mainly determined by the thermal and magnetic characteristics of the circuit breakers. The upstream and downstream time-current characteristics are chosen in such a way that the downstream circuit breaker clears the fault first, and the upstream circuit breaker only clears the fault when the downstream circuit breaker fails.

It was shown in the previous section that the commutation of inductive currents can cause the undesired tripping of SSCBs in non-faulted sections. Therefore, coordination among downstream circuit breakers is also required, not just between upstream and downstream SSCBs. Furthermore, since this paper aims for a plug-and-play protection scheme, the coordination must also achieve selectivity in meshed low voltage dc grids.

To prevent unnecessary tripping due to commutated currents, the time-current current characteristic must take this current and its decay into account. In the worst-case, the commutated current is the nominal current and this current decays with the time constant of the line. If the SSCB is carrying the nominal current before commutation, the worst-case current after commutation is characterized by the LR time constant and is given by

$$I_{total} = I_{nom}\left(1 + e^{-\frac{R_L}{L_L}t}\right). \tag{9}$$

Therefore, in order to prevent the SSCB from tripping unnecessarily from commutated inductive currents, the proposed characteristic only interrupts immediately if the current exceeds twice the nominal current. Furthermore, between I_{nom} and $2I_{nom}$, the time-current characteristic is chosen as

$$t_{clear} = t_{max} - \frac{L_L}{R_L}\ln(I+1), \tag{10}$$

where t_{max} is the maximum time the SSCB takes to detect and clear an overcurrent, and I is the current in the SSCB in multiple of the nominal current.

Note that it is only necessary to know the slowest expected time constant of the lines in the system. Therefore, knowledge about the length of the lines in the system or their interconnection is not required. Moreover, a safety margin can be implemented in order to anticipate uncertainty in the system parameters. The proposed time-current characteristic for SSCBs is shown in Figure 17.

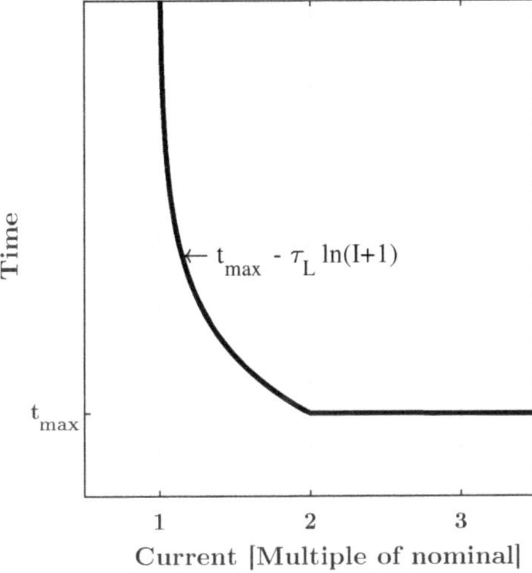

Figure 17. Proposed time-current characteristic for the plug-and-play protection scheme, where t_{max} is the maximum time that the SSCB needs to detect and interrupt an overcurrent.

Because the time-current characteristic scales with nominal current, the proposed time-current characteristic inherently coordinates upstream and downstream SSCBs. To illustrate this, imagine the system of Figure 12 with an upstream SSCB that has a nominal current of 32 A and two downstream SSCBs that have a nominal current of 16 A. Now if all the SSCBs operate at their nominal current and a fault occurs downstream, the downstream SSCB will trip immediately when the current reaches 32 A. In this case the upstream current is 48 A, for which the upstream breaker will wait a significant time.

5.3. Plug-and-Play Design Guidelines

In the previous subsections a solid-state circuit breaker topology and time-current characteristic were proposed in order to ensure selectivity. In this subsection it is described how these concepts can be incorporated in an SSCB, with a nominal current of I_{nom} and a nominal voltage of U_{nom}, in order to achieve system-wide plug-and-play protection selectivity.

The SSCB's di/dt detection is tripped if the voltage over the current limiting inductance is more than $U_{L,max}$. However, in the worst case the voltage is still U_{nom} for t_{max}. Therefore, assuming a sawtooth shaped pulse, the required current limiting inductance is determined by

$$L_{CB} = \frac{\sqrt{3}U_{nom}t_{max}}{I_{pulse}(t_{max})}, \tag{11}$$

where $I_{pulse}(t_{max})$ is the current carrying capability of the semiconductor switches for a t_{max} pulse, which usually is several times higher than then nominal current of the switches.

When the current is above two times the nominal current, the clearing time is given by t_{max}, while for currents between one and two times the nominal current (10) is used to determine the clearing time. Consequently, the maximum current when the overcurrent protection is tripped is given by

$$I_{max} = 2I_{nom} + \frac{U_{L,max}t_{max}}{L_{CB}}. \tag{12}$$

Although lowering the di/dt threshold $U_{L,max}$ decreases the maximum fault current, the detection will also become more sensitive to, for example, electromagnetic interference. The authors found a reasonable threshold voltage to be around

$$U_{L,max} = \frac{2I_{nom}L_{CB}}{t_{max}}. \tag{13}$$

In general, changes in load current will not trigger the di/dt detection with this threshold, since the time constants of distribution lines and power electronic converters are several orders of magnitude higher than the time constant of the SSCB.

The topology that is used for the SSCB is shown in Figure 14, where the MOVs clamp the voltage to below the maximum rating of the switches. Alternatively, other circuits can be used to limit the voltage on the switches. Regardless, the rating of the clamping circuits determines the maximum inductive energy that the SSCBs can dissipate.

To size the damper components (7) and (8) are used. To ensure a smooth commutation of inductive current and prevent instant fault propagation, the resonant frequency of the SSCB is chosen to be an order of magnitude lower than the inverse of the maximum clearing time t_{max} (in this paper a factor of 10 is chosen). The damper capacitor is then given by

$$C_d \gg \frac{t_{max}^2}{4\pi^2 L_{CB}}. \tag{14}$$

For a damped response, the damper resistance is sized such that the attenuation frequency is higher than the resonant frequency. Therefore,

$$R_d > 2\sqrt{\frac{L_{CB}}{C_d}}. \tag{15}$$

If possible, in order to clamp the voltage over the inductor to below the threshold voltage during commutation, the damper resistance should also be

$$R_d < \frac{U_{L,max}}{I_{nom}}. \tag{16}$$

Utilizing these guidelines, the parameters of the SSCB in this paper are given in Table 2. As a consequence of the damper capacitance and damper resistance, the resonant frequency f_r of the SSCB is 113 kHz and the attenuation frequency α is 170 kHz. Furthermore, the maximum steady-state losses in the SSCB are 26 W, which is only around 0.7% of the conducted power. Moreover, the losses in the SSCB can be reduced even further by, for example, parallelling multiple semiconductors.

Table 2. Design parameters of the solid-state circuit breaker.

Parameter	Acronym	Value
Nominal voltage	U_{nom}	350 V
Nominal current	I_{nom}	10 A
On-state resistance per pole	R_{CB}	130 mΩ
Current limiting inductance	L_{CB}	1.0 μH
Damper resistance	R_d	2.0 Ω
Damper capacitance	C_d	2.0 μF
Minimum clearing time	t_{max}	1.0 μs
Overcurrent threshold	$I_{L,max}$	20 A
di/dt threshold	$U_{L,max}$	20 V

6. Experimental Validation of the Plug-and-Play Protection Scheme

To show that the plug-and-play SSCBs delay the propagation of the fault, the experiment shown in Figure 10 is repeated. The experimental results for the currents in the SSCBs for this experiment are shown in Figure 18. Observe that, contrary to the experiment in Figure 11, only CB3 is tripped, while the currents in the other SSCBs are largely unaffected by the whole process. The current rises fast until the switches of CB3 are opened, after which CB1 and CB2 remain closed. Furthermore, it is seen that the oscillations in the system are attenuated significantly because of the RC dampers.

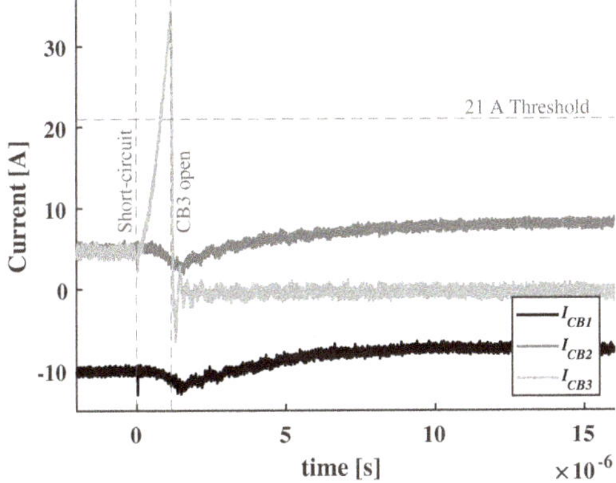

Figure 18. Experimental results for the system shown in Figure 10, showing that fault propagation is delayed with the plug-and-play SSCBs.

To show that the plug-and-play SSCBs ensure smooth commutation and selectivity, the experiment shown in Figure 12 is repeated. The experimental results for the SSCBs' currents for this experiment are shown in Figure 19. Note that the commutation of the inductive current is smoothed out over roughly a 10 μs interval, which is an order of magnitude longer than in Figure 13. Furthermore, although the inductive current is commutated to CB1, its thresholds are not exceeded and therefore its fault detection is not tripped.

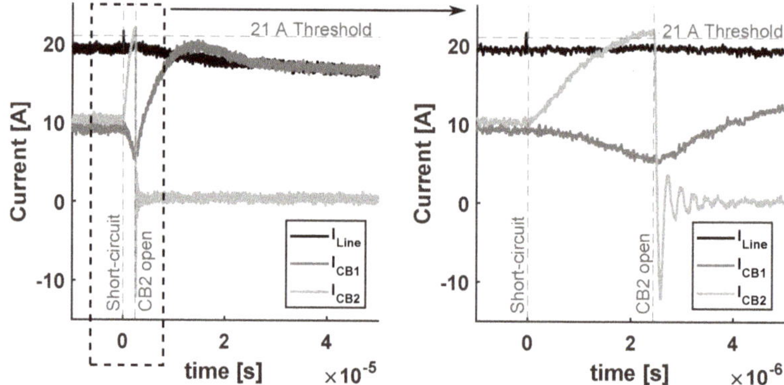

Figure 19. Experimental results for the system shown in Figure 12, which show smooth commutation and selectivity with the plug-and-play SSCBs (the right figure is a zoom in).

From the previous experiment, it is clear that the plug-and-play protection scheme accounts for commutated currents, but does not prevent them. The addition of a significant capacitance at the interface of the SSCBs can reduce the commutated current by (temporarily) storing the inductive energy. The experimental results for the same experiment, but with an added 240 µF capacitance at the interface of the SSCBs, is shown in Figure 20. The commutated current is reduced, but applying this solution in a plug-and-play fashion is impractical, since information about the system's capacitances and inductances is required. Therefore, this paper adopted the proposed time-current characteristic.

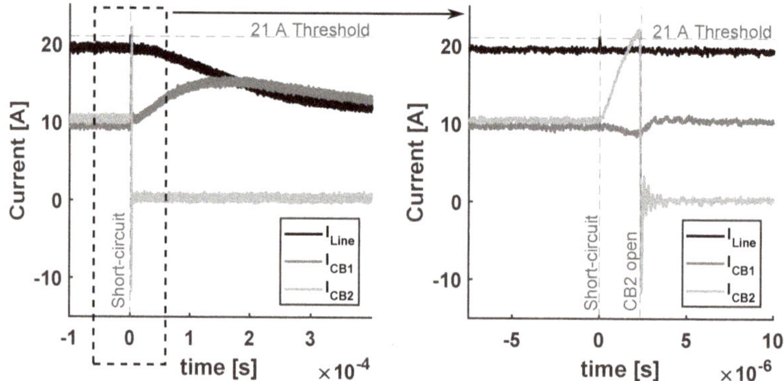

Figure 20. Experimental results for the system shown in Figure 12 when a capacitance of 240 µF is added at the interface of the plug-and-play SSCBs, showing that the commutated current is reduced (the right figure is a zoom in).

To show that the decentralized plug-and-play protection scheme also works for meshed systems, the experiment shown in Figure 21 is used. The setup consists of a constant voltage source of 350 V connected to two 5 A constant current loads in a ring configuration. The lines in this system are emulated by equivalent π-circuits described in Section 3. This situation can occur, for example, when dc households are interconnected.

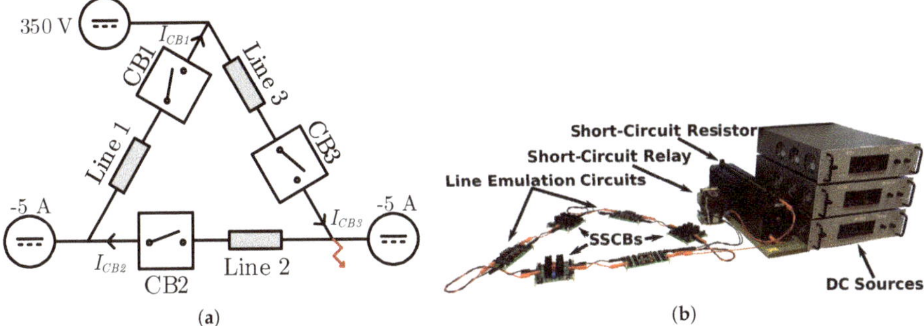

Figure 21. (a) Schematic and (b) picture of the experimental setup connecting a constant voltage source to two constant current loads via lines in a meshed configuration.

The experimental results for the currents inside the SSCBs, when a short-circuit with a fault resistance of 2.5 Ω is induced at the load-side terminals of CB3, are given in Figure 22. Observe that the fault current is first mostly supplied by the damper capacitances of CB3 because this path contains the lowest inductance, and consequently CB3's di/dt detection is triggered. Subsequently, the fault current is supplied mainly by the capacitance of the load in the non-faulted part of the system through CB2 and its overcurrent detection is triggered after around 20 µs. Most importantly, CB1 is not triggered and the non-faulted section of the grid remains operational.

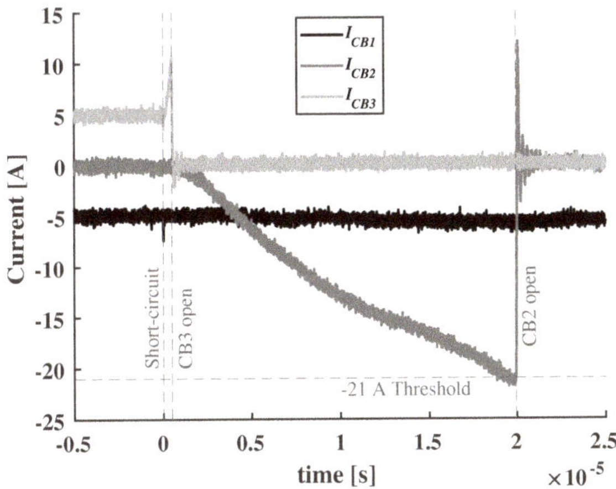

Figure 22. Experimental results for the system shown in Figure 21, showing that selectivity is also achieved in meshed systems.

Discussion

The experimental results show that the proposed decentralized plug-and-play protection scheme ensures selectivity for different dc grids under various conditions. Firstly, the systems were tested with varying impedance between the sources and the faults. In the experiments, short-circuits with low fault impedance generally trip the di/dt detection, while short-circuits with high fault impedance generally trip the overcurrent detection. Secondly, the experiments showed that the SSCBs are capable of interrupting non-inductive and inductive fault currents, dissipating the inductive energy in the MOV's when there is no alternative path for inductive currents. Thirdly, it is shown that the protection scheme achieves this for both radial and meshed systems.

The main advantages of the proposed protection scheme are that no communication infrastructure is required and that only minimal knowledge about the system is needed. Furthermore, because the SSCBs are designed for the worst-case scenario, the approach is not sensitive to uncertainties and disturbances in the system. Moreover, the protection scheme is resilient to failures, since upstream circuit breakers trip if downstream circuit breakers fail. Additionally, it is important to note that the RC dampers do not increase the steady state losses of the SSCBs, although they slightly increase losses during transient situations.

The main disadvantage of the proposed protection scheme is that the components in the system need to sustain twice the nominal current for up to $t_{\max}$. Furthermore, the worst-case time constant of the lines in the system need to be known, although this is not influenced by the length of the lines or their configuration. Moreover, care must be taken that the normal behavior of sources and loads, such as inrush currents, does not cause unnecessary tripping. However, the time-current characteristic can be adapted to avoid tripping on these events. Additionally, although solid-state circuit breakers solve a lot of challenges with regards to the protection of dc grids, they are still relatively inefficient and expensive compared to mechanical circuit breakers.

For the industrial application of the proposed plug-and-play SSCBs, different design considerations can be made. For example, the SSCB's current can be measured utilizing the voltage over the switches instead of using a high-bandwidth hall-sensor, or the di/dt measurement can be omitted in order to reduce complexity. Furthermore, semiconductor switches with lower cost or lower on-state resistance can be used, or multiple switches can be put in parallel or series. Overall, trade-offs can be made for the electrical performance of the SSCB versus its cost. However, detailed cost analysis and optimization of SSCBs were outside of the scope of this paper.

The proposed plug-and-play protection scheme provides a solid foundation for the protection of low voltage dc systems. In future research, the scheme's applicability to systems with longer distribution lines, such as medium and high voltage transmission systems, can be investigated. In those studies the impact of the propagation delay on the effectiveness of the approach must be examined. Furthermore, more cost and energy efficient solid-state circuit breaker topologies must be explored. Moreover, it can be researched how the SSCBs can provide additional functionality, such as controlling the power flow and black-starting dc grids. Additionally, a communication infrastructure can be used to improve performance or add functionality.

7. Conclusions

The lack of a zero-crossing makes interrupting inductive currents more challenging in dc grids than in conventional ac grids. Furthermore, fast fault interruption is often required for low voltage dc grids in order to reduce the current stress on the components in the grid and prevent blackouts. Moreover, meshed topologies and bi-directional power flow complicate fault detection and selectivity. To ensure selectivity, literature presents several schemes that rely on communication, or knowledge about the system's topology and parameters. However, fast fault interruption is difficult when communication is used, and systems are subject to uncertainty and change.

Because of the fast fault interruption and meshed system structures, it is difficult to ensure protection selectivity in low voltage dc grids with non-unit protection schemes. It was experimentally shown that the current and current rate of change of circuit breakers in non-faulted regions of the grid can exceed their thresholds before the faulted circuit breakers can clear the fault. Furthermore, interrupted inductive currents temporarily commutate to healthy parts of the system, causing overcurrents. As a consequence of these two phenomena, undesired tripping can occur of circuit breakers in non-faulted parts of the system.

This paper proposes a decentralized plug-and-play protection scheme that ensures security and selectivity, without utilizing communication and with minimal knowledge of the system. The protection scheme delays fault propagation by introducing RC dampers at both ends of the solid-state circuit breaker, which then forms a second order filter for the fault. Furthermore, commutated inductive overcurrents are ignored by incorporating the lines' worst time constant into a time-current characteristic. Additionally, design guidelines were provided for the proposed solid-state circuit breaker topology. Several experiments were carried out that showed that the proposed protection scheme provides secure and selective fault interruption for radial and meshed low voltage dc grids.

Although the proposed protection scheme provides an effective and robust solution, further research is still required. The effectiveness of the protection scheme for medium and high voltage grids must be investigated. Moreover, more research is required on solid-state protection devices, since they are still inefficient and expensive compared to their mechanical counterparts. Additionally, SSCBs have potential for providing additional functionality to the grid, with or without utilizing a communication infrastructure.

Author Contributions: Conceptualization, N.H.v.d.B. and P.P.; methodology, N.H.v.d.B. and P.P.; validation, N.H.v.d.B. and T.B.S.; formal analysis, N.H.v.d.B.; investigation, N.H.v.d.B. and T.B.S.; resources, T.B.S.; writing–original draft preparation, N.H.v.d.B.; writing–review and editing, L.M.R.-E., M.T.J.S. and T.B.S.; visualization, N.H.v.d.B.; supervision, L.M.R.-E., M.T.J.S. and T.B.S.; project administration, P.B.; funding acquisition, P.B. All authors have read and agreed to the published version of the manuscript.

Funding: This project has received funding in the framework of the joint programming initiative ERA-Net Smart Grids Plus, with support from the European Union's Horizon 2020 research and innovation programme.

Conflicts of Interest: The authors declare no conflict of interest. Furthermore, the funders had no role in the design of the study; in the collection, analyses, or interpretation of data; in the writing of the manuscript, or in the decision to publish the results.

References

1. Hakala, T.; Lahdeaho, T.; Jarventausta, P. Low-Voltage DC Distribution—Utilization Potential in a Large Distribution Network Company. *IEEE Trans. Power Deliv.* **2015**, *30*, 1694–1701. [CrossRef]
2. Allee, G.; Tschudi, W. Edison Redux: 380 Vdc Brings Reliability and Efficiency to Sustainable Data Centers. *IEEE Power Energy Mag.* **2012**, *10*, 50–59. [CrossRef]
3. Larruskain, D.M.; Zamora, I.; Abarrategui, O.; Aginako, Z. Conversion of AC distribution lines into DC lines to upgrade transmission capacity. *Electr. Power Syst. Res.* **2011**, *81*, 1341–1348. [CrossRef]
4. Guerrero, J.M.; Vasquez, J.C.; Matas, J.; de Vicuna, L.G.; Castilla, M. Hierarchical Control of Droop-Controlled AC and DC Microgrids: A General Approach Toward Standardization. *IEEE Trans. Ind. Electron.* **2011**, *58*, 158–172. [CrossRef]
5. Arcidiacono, V.; Monti, A.; Sulligoi, G. Generation control system for improving design and stability of medium-voltage DC power systems on ships. *IET Electr. Syst. Transp.* **2012**, *2*, 158–167. [CrossRef]
6. Gregory, G.D. Applying low-voltage circuit breakers in direct current systems. *IEEE Trans. Ind. Appl.* **1995**, *31*, 650–657. [CrossRef]
7. Dragicevic, T.; Lu, X.; Vasquez, J.C.; Guerrero, J.M. DC Microgrids—Part II: A Review of Power Architectures, Applications, and Standardization Issues. *IEEE Trans. Power Electron.* **2016**, *31*, 3528–3549. [CrossRef]
8. Brearley, B.J.; Prabu, R.R. A review on issues and approaches for microgrid protection. *Renew. Sustain. Energy Rev.* **2017**, *67*, 988–997. [CrossRef]
9. Hailu, T.; Mackay, L.; Gajic, M.; Ferreira, J.A. Protection coordination of voltage weak DC distribution grid: Concepts. In Proceedings of the IEEE 2nd Annual Southern Power Electronics Conference (SPEC), Auckland, New Zealand, 5–8 December 2016.
10. Shen, Z.J. Ultrafast Solid-State Circuit Breakers: Protecting Converter-Based ac and dc Microgrids Against Short Circuit Faults. *IEEE Electrif. Mag.* **2016**, *4*, 60–70. [CrossRef]
11. Guillod, T.; Krismer, F.; Kolar, J.W. Protection of MV Converters in the Grid: The Case of MV/LV Solid-State Transformers. *IEEE J. Emerg. Sel. Top. Power Electron.* **2017**, *5*, 393–408. [CrossRef]
12. Qi, L.; Antoniazzi, A.; Raciti, L. DC Distribution Fault Analysis, Protection Solutions, and Example Implementations. *IEEE Trans. Ind. Appl.* **2018**, *54*, 3179–3186. [CrossRef]
13. Brozek, J.P. DC overcurrent protection-where we stand. *IEEE Trans. Ind. Appl.* **1993**, *29*, 1029–1032. [CrossRef]

14. ABB. Circuit-Breakers for Direct Current Applications. Available online: https://library.e.abb.com/public/de4ebee4798b6724852576be007b74d4/1SXU210206G0201.pdf (accessed on 17 June 2020).

15. Lazzari, R.; Piegari, L. Design and Implementation of LVDC Hybrid Circuit Breaker. *IEEE Trans. Power Electron.* **2019**, *34*, 7369–7380. [CrossRef]

16. Sato, Y.; Tanaka, Y.; Fukui, A.; Yamasaki, M.; Ohashi, H. SiC-SIT Circuit Breakers With Controllable Interruption Voltage for 400-V DC Distribution Systems. *IEEE Trans. Power Electron.* **2014**, *29*, 2597–2605. [CrossRef]

17. Miao, Z.; Sabui, G.; Chen, A.; Li, Y.; Shen, Z.J.; Wang, J.; Shuai, Z.; Luo, A.; Yin, X.; Jiang, M. A self-powered ultra-fast DC solid state circuit breaker using a normally-on SiC JFET. In Proceedings of the IEEE Applied Power Electronics Conference and Exposition (APEC), Charlotte, NC, USA, 15–19 March 2015; pp. 767–773.

18. Bayati, N.; Hajizadeh, A.; Soltani, M. Protection in DC microgrids: a comparative review. *IET Smart Grid* **2018**, *1*, 66–75. [CrossRef]

19. Mirsaeidi, S.; Said, D.; Mustafa, M.; Habibuddin, M.; Miveh, M. A Comprehensive Overview of Different Protection Schemes in Micro-Grids. *Int. J. Emerg. Electr. Power Syst. (IJEEPS)* **2013**, *14*, 327–332. [CrossRef]

20. Meng, L.; Shafiee, Q.; Trecate, G.F.; Karimi, H.; Fulwani, D.; Lu, X.; Guerrero, J.M. Review on Control of DC Microgrids and Multiple Microgrid Clusters. *IEEE J. Emerg. Sel. Top. Power Electron.* **2017**, *5*, 928–948.

21. Salomonsson, D.; Soder, L.; Sannino, A. Protection of Low-Voltage DC Microgrids. *IEEE Trans. Power Deliv.* **2009**, *24*, 1045–1053. [CrossRef]

22. Meghwani, A.; Srivastava, S.C.; Chakrabarti, S. A Non-unit Protection Scheme for DC Microgrid Based on Local Measurements. *IEEE Trans. Power Deliv.* **2017**, *32*, 172–181. [CrossRef]

23. Sneath, J.; Rajapakse, A.D. Fault Detection and Interruption in an Earthed HVDC Grid Using ROCOV and Hybrid DC Breakers. *IEEE Trans. Power Deliv.* **2016**, *31*, 973–981. [CrossRef]

24. Yang, J.; Fletcher, J.E.; O'Reilly, J. Short-Circuit and Ground Fault Analyses and Location in VSC-Based DC Network Cables. *IEEE Trans. Ind. Electron.* **2012**, *59*, 3827–3837. [CrossRef]

25. Mohanty, R.; Pradhan, A.K. A Superimposed Current Based Unit Protection Scheme for DC Microgrid. *IEEE Trans. Smart Grid* **2018**, *9*, 3917–3919. [CrossRef]

26. Bertho, R.; Lacerda, V.A.; Monaro, R.M.; Vieira, J.C.M.; Coury, D.V. Selective Nonunit Protection Technique for Multiterminal VSC-HVDC Grids. *IEEE Trans. Power Deliv.* **2018**, *33*, 2106–2114. [CrossRef]

27. Som, S.; Samantaray, S.R. Efficient protection scheme for low-voltage DC micro-grid. *IET Gener. Transm. Distrib.* **2018**, *12*, 3322–3329. [CrossRef]

28. Feng, X.; Qi, L.; Pan, J. A Novel Fault Location Method and Algorithm for DC Distribution Protection. *IEEE Trans. Ind. Appl.* **2017**, *53*, 1834–1840. [CrossRef]

29. Tang, L.; Ooi, B. Locating and Isolating DC Faults in Multi-Terminal DC Systems. *IEEE Trans. Power Deliv.* **2007**, *22*, 1877–1884. [CrossRef]

30. Emhemed, A.A.S.; Burt, G.M. An Advanced Protection Scheme for Enabling an LVDC Last Mile Distribution Network. *IEEE Trans. Smart Grid* **2014**, *5*, 2602–2609. [CrossRef]

31. Fletcher, S.D.A.; Norman, P.J.; Fong, K.; Galloway, S.J.; Burt, G.M. High-Speed Differential Protection for Smart DC Distribution Systems. *IEEE Trans. Smart Grid* **2014**, *5*, 2610–2617. [CrossRef]

32. Farhadi, M.; Mohammed, O.A. Event-Based Protection Scheme for a Multiterminal Hybrid DC Power System. *IEEE Trans. Smart Grid* **2015**, *6*, 1658–1669. [CrossRef]

33. Farhadi, M.; Mohammed, O.A. A New Protection Scheme for Multi-Bus DC Power Systems Using an Event Classification Approach. *IEEE Trans. Ind. Appl.* **2016**, *52*, 2834–2842. [CrossRef]

34. Emhemed, A.A.S.; Fong, K.; Fletcher, S.; Burt, G.M. Validation of Fast and Selective Protection Scheme for an LVDC Distribution Network. *IEEE Trans. Power Deliv.* **2017**, *32*, 1432–1440. [CrossRef]

35. Monadi, M.; Gavriluta, C.; Luna, A.; Candela, J.I.; Rodriguez, P. Centralized Protection Strategy for Medium Voltage DC Microgrids. *IEEE Trans. Power Deliv.* **2017**, *32*, 430–440. [CrossRef]

36. Park, J.; Candelaria, J. Fault Detection and Isolation in Low-Voltage DC-Bus Microgrid System. *IEEE Trans. Power Deliv.* **2013**, *28*, 779–787. [CrossRef]

37. Wang, L. The Fault Causes of Overhead Lines in Distribution Network. *MATEC Web Conf.* **2016**, *61*, 02017. [CrossRef]
38. Paul, C.R. *Analysis of Multiconductor Transmission Lines*; Wiley-IEEE Press: Hoboken, NJ, USA, 2007; ISBN 0470131543.

Article

A Local Protection and Local Action Strategy of DC Grid Fault Protection

Jingqiu Yu, Zheren Zhang and Zheng Xu *

Department of Electrical Engineering, Zhejiang University, Hangzhou 310058, China;
yujingqiu@zju.edu.cn (J.Y.); 3071001296zhang@zju.edu.cn (Z.Z.)
* Correspondence: xuzheng007@zju.edu.cn

Received: 24 August 2020; Accepted: 12 September 2020; Published: 14 September 2020

Abstract: Fast detection and isolation of direct current (DC) faults are key issues for DC grids. Therefore, it is very necessary to study the fault protection principle for DC grids. This paper firstly presents the main difficulties in DC fault protection. Then, a local protection and local action strategy for isolating the DC faults is proposed. To illustrate the performance of the proposed protection strategy, a four-terminal DC grid with the hybrid high voltage direct current (HVDC) circuit breakers (HVDC CBs) is constructed in the time-domain simulation software PSCAD/EMTDC as the test system. The systematical comparison between the ordinary protection strategy and the proposed strategy is carried out. The protection selectivity of the proposed local detection and local action strategy is thoroughly studied through complete DC line fault scanning of the test system. The simulation results show that the proposed strategy is of high protection selectivity and speed. In addition, the current rating and the voltage of HVDC CB could be greatly reduced with the proposed strategy, which proves the economic benefits of the proposed strategy.

Keywords: DC grids; fault protection; local detection; local action; DC circuit breaker

1. Introduction

Theoretically, the converters in the DC grid can be connected in parallel, in series, or in hybrid mode [1]. For the parallel converters, the DC voltage of each converter is roughly the same, and the power distribution can be achieved by controlling the DC current of each converter. For converters connected in series, the current flowing through each converter is the same, and the power distribution can be regulated by adjusting the DC voltage of each converter. However, the converters connected in parallel are more suitable for DC grids due to the higher reliability, economy, and flexibility. In this configuration, the voltage polarity of each converter is fixed. In the early ages, most high voltage direct current (HVDC) systems were based on the line commutated converters (LCCs), and the DC power flow direction was fixed since the current of the thyristor could not be reversed. Due to this shortcoming in LCC, DC grid has not received much attention in the first 50 years of the HVDC transmission. However, with the development of power electronics, the voltage source converter-based HVDC (VSC-HVDC) [2–5] has drawn much attention from both academia and industry due to its flexible operation capability, and the advantages of a DC grid can be fully achieved by VSC.Ever since, the DC grid has become a new expectation of the power industry.

There are three main technical bottlenecks in the development of DC grids. One is the fast detection and isolation technology of DC faults [6,7]. The second is the DC voltage transformation technology [8,9]. The third is the power flow control technology of the DC lines [10,11]. This paper mainly deals with the first technical bottleneck and studies the fault protection principle for DC grids.

The difficulties of DC grids in fault protection lie mainly in the following aspects [12–15]:

(1) The fault current rises very quickly. Generally, the fault current could reach its steady value within 10 ms after the fault;

(2)   The steady fault current is very large, and it could reach dozens of times the rated current;

(3)   There is no zero-crossing point in the fault current, which makes it difficult for the circuit breakers (CBs) to extinguish the fault arc;

(4)   The requirements to fast clear the fault current are extremely high. The fault isolation time in the alternating current (AC) system is generally 50 ms and above, while it is required to clear the DC faults in 5 ms in the DC grid. Otherwise, the equipment safety will be threatened. To achieve the rapidity of DC fault protection in the DC grid, the DC faults are supposed not to be cleared mainly through the relay protection at the AC side.

To this end, this paper proposes a local protection and local action strategy for HVDC grids. This strategy is of high protection selectivity and speed, which is able to solve the above technical problems effectively.

This paper is organized as follows. Section 2 gives the basic structure and operation principle of the hybrid circuit breaker. The local protection and local action strategy is introduced in Section 3. Section 4 describes the test system constructed in this paper. To verify the improved protection speed and selectivity, a performance comparison between the proposed strategy and the conventional relay protection strategy is presented in Section 5. The conclusions are drawn in Section 6.

2. Basic Structure and Operation Principle of DC Circuit Breaker

The HVDC CB adopted in this paper is the hybrid HVDC CB proposed by ABB [16,17], as shown in Figure 1. The HVDC CB is connected between one multi-level modular converter (MMC) and one DC line. When the faults occur on the DC lines, the HVDC CB starts action to isolate the fault line.

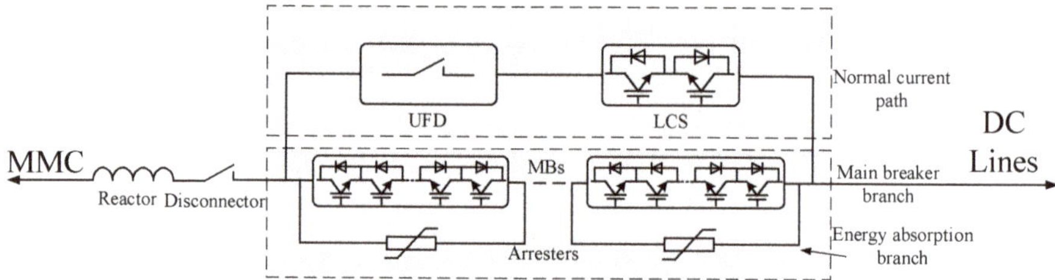

Figure 1. Structure of the hybrid high voltage direct current circuit breaker (HVDC CB).

The DC circuit breaker is composed of a normal current path, a main breaker branch, and an energy absorption branch. The normal current path consists of a load commutation switch (LCS) and an ultra-fast disconnector (UFD) connected in series. Multiple main breakers (MBs) connected in series constitute the main breaker branch. The basic structure of each part is as follows:

(1)   Main breakers (MBs). This is the core component of HVDC CB. It determines the voltage withstand capacity and current breaking capacity. Furthermore, it is composed of multiple parts connected in series. Each part involves multiple semiconductor switches connected in series, together with arresters in parallel. To improve the breaking capacity, a single semiconductor switch can be composed of multiple insulated-gate bipolar transistors (IGBTs) in parallel. MBs need to have bidirectional current interruption capability and withstand the pole-to-ground voltage;

(2)   Arresters. Each individual main breaker module is in parallel with the arrester banks. The arresters absorb energy and make the fault current decay quickly;

(3)   Load commutation switch (LCS). Its structure is similar to the segmented structure of MBs. It is composed of multiple switches connected in series, which includes several IGBTs connected in series and the anti-parallel diodes. Since LCS does not need to withstand high voltages, only a few power electronic devices are required. Moreover, LCS needs to have bidirectional current breaking capability;

(4)   Ultra-fast disconnector (UFD). It needs to quickly disconnect the circuit under zero current state. Furthermore, the breaking time is about 2 ms.

The principle of the hybrid HVDC CB is as follows [18–20]:

(1)   During the steady state, UFD and LCS are closed, whereas MBs are open. The DC current flows through the normal current path;

(2)   When a fault occurs on the DC side, the IGBTs in LCS start action, and then the fault current transfers to the fault current interruption branch after a delay of about 250 µs;

(3)   When the current in LCS drops to the residual current, UFD is commanded to open. Moreover, it achieves full contact separation within 2 ms;

(4)   After UFD finishes its operation, MBs start action. At the moment MBs are disconnected, the arresters are inserted to the fault line. This causes a sudden change in the flowing path of the fault current. Due to the sudden change of the inductor current, there would be an extremely high overvoltage in the DC grid (which could exceed twice the rated DC voltage). The insertion of the arresters makes the fault current attenuate to zero. Usually, it takes 10–15 ms for the fault current to decay to zero.

3. Two Basic Protection Strategies for DC Grids

For a half-bridge sub-module (HBSM) multi-level modular converters (MMCs)-based DC grid equipped with HVDC CBs, how to quickly detect and isolate the DC faults is a very challenging problem.

The conventional strategy is to follow the practice of the AC grid. The protection action sequence is shown in Figure 2. Firstly, the relay protection system detects the fault location, and then the fault line is isolated by CBs. However, this strategy puts extremely high requirements on the speed and selectivity of the relay protection system. According to the speed of fault detection in conventional two-terminal HVDC systems, the duration for fault detection is about 10 ms [21]. If the fault detection speed in DC grids is the same as that in the conventional two terminal HVDC system, the fault current to be interrupted by the DC breaker will reach a very high level. Furthermore, this will result in huge costs of CBs. In addition, this will severely limit the application of DC grids of this structure.

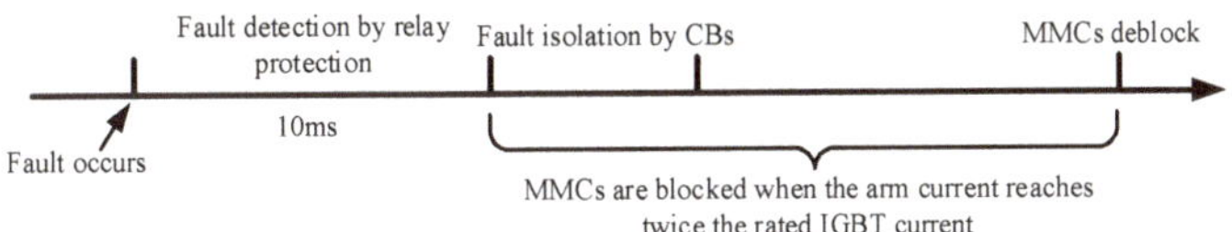

Figure 2. Protection action sequence of strategy 1.

The other strategy is based on local protection and local action. There could be two aspects to explain local protection. The first is the local protection of the converters. If the arm current reaches twice the rated IGBT current, the converter will then be automatically blocked. Moreover, there is no need for a relay protection signal. The second means the local protection of CBs. When the current flowing through CB is more than twice its rated current, LCS will be activated immediately as well as the whole CB. That is, CBs on both sides of the line independently finish fault detection and tripping, and there is no need for a coordination between CBs. Local action means that CB only operates when the fault occurs at the DC line where CB is located. Define that the positive direction of the CB current is from MMC to the DC line. Then, only when the fault current is the same as the positive direction and reaches twice the rated CB current, CB is activated. Local action guarantees the high protection selectivity. The protection action sequence is shown in Figure 3. Practice has shown that this strategy is very suitable for DC grids. Furthermore, it has high protection speed and protection selectivity, which could greatly reduce the required breaking capacity of CBs, thereby saving the cost.

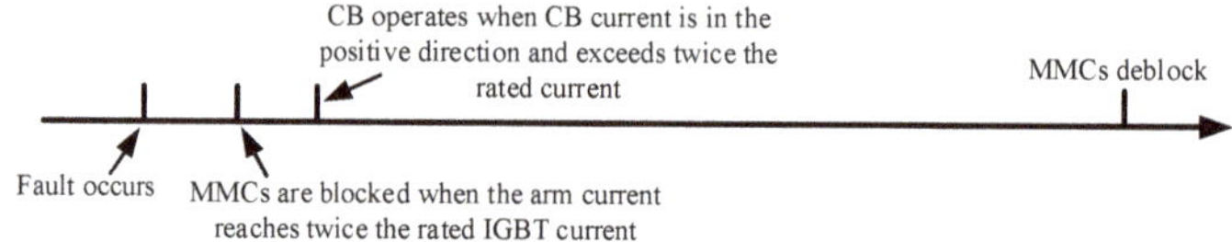

Figure 3. Protection action sequence of strategy 2.

The two different strategies will be studied separately in the following sections. The first strategy is to deal with the faults by the conventional DC fault relay protection. The second strategy is the local protection and local action strategy.

4. Test System

The two different strategies will be compared through simulations of a test system on PSCAD/EMTDC. The designed test system is shown in Figure 4. Both the AC system at the sending end and the receiving end are connected to a modified two-zone four-generator system [22]. The main parameters of the four-machine system are listed in Table 1. The generator and the control system parameters, as well as the network structure and parameters remain consistent with the original system. Only the load and output of the generator are changed. The DC grid is in the monopole-grounded return operation mode. The model of all DC lines is $4 \times$ LGJ-720. The parameters of DC lines in the simulation are resistance 0.009735 Ω/km, inductance 0.8489×10^{-3} H/km, and capacitance 0.01367×10^{-6} F/km. The parameters of the four MMCs based on HBSMs are shown in Table 2. The control mode is shown in Table 3. Table 4 gives the initial state of the two AC systems. Among them, G_{A1}–G_{A4} and G_{B1}–G_{B4} are the generators of the two AC systems A and B, respectively. L_{A7}, L_{A9} and L_{B7}, L_{B9} are the loads connected to the No. 7 and No. 9 buses. C_{A7}, C_{A9} and C_{B7}, C_{B9} are the reactive power compensators.

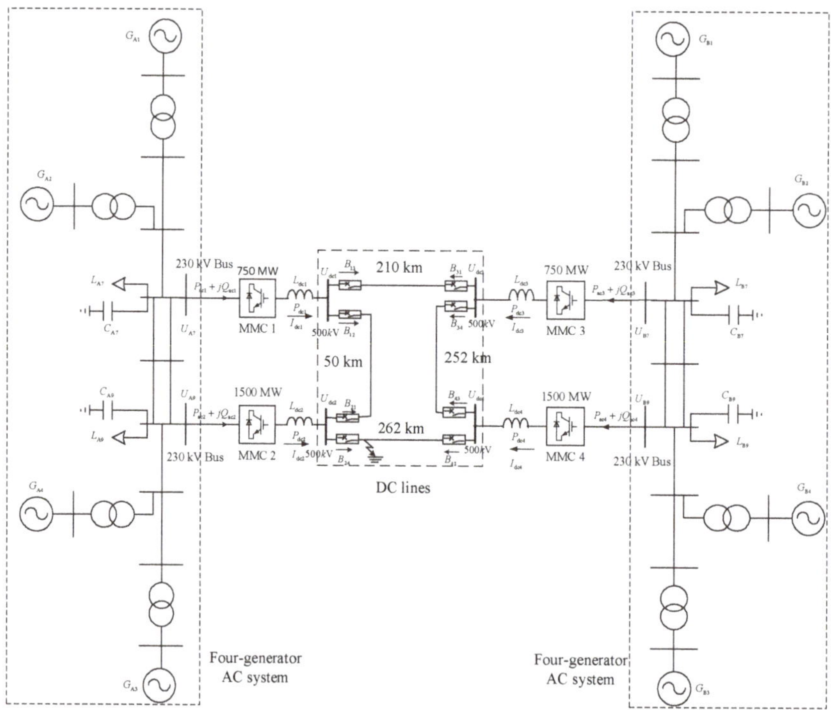

Figure 4. Structure of the designed test system.

Table 1. Main parameters of the four-generator alternating current (AC) system.

Item	Values
Generator G1-G4	
Rated RMS voltage	20 kV
Rated 3 Phase MVA	900 MVA
T_{d0}'	8.0 s
T_{q0}'	0.4 s
T_{d0}''	0.25 s
T_{q0}''	0.25 s
X_d	7.2 Ω
X_q	6.8 Ω
X_d'	1.2 Ω
X_q'	2.2 Ω
X_d''	1.0 Ω
X_q''	1.0 Ω
Transformer	
Transformer ratio	20 kV/230 kV
Positive sequence leakage reactance	8.82 Ω
AC lines	
Length	100 km
+ve sequence R	1.32×10^{-2} Ω/km
+ve sequence L	1.32×10^{-1} H/km
+ve sequence C	8.28×10^{-7} F/km

Table 2. Main parameters of multi-level modular converters (MMCs) in the test system.

	MMC 1	MMC 2	MMC 3	MMC 4
Rated capacity/MVA	750	1500	750	1500
AC bus voltage/kV	230	230	230	230
Rated DC voltage/kV	500	500	500	500
Transformer rated capacity/MVA	900	1800	900	1800
Transformer nominal ratio	230/255	230/255	230/255	230/255
Transformer inductance/%	15	15	15	15
SM rated voltage/kV	1.6	2.2	1.6	2.2
Number of SMs per arm	313	228	313	228
SM capacitance/mF	12	18	12	18
Arm inductance/mH	66	32	66	32
Smoothing reactor/mH	300	300	300	300

Table 3. Control mode and reference value of MMCs.

MMC	Control Mode	Reference Value
1	d-axis: Constant active power control q-axis: Constant reactive power control	$P_{dc1} = 200$ MW; $Q_{ac1} = 0$ Mvar
2	d-axis: Constant active power control q-axis: Constant reactive power control	$P_{dc2} = 400$ MW; $Q_{ac2} = 0$ Mvar
3	d-axis: Constant active power control q-axis: Constant reactive power control	$P_{dc3} = -200$ MW; $Q_{ac3} = 0$ Mvar
4	d-axis: Constant voltage control q-axis: Constant reactive power control	$U_{dc4} = 500$ kV; $Q_{ac4} = 0$ Mvar

Table 4. Initial state of the generators, loads, and reactive power compensators in the AC systems.

	Active Power/MW	Reactive Power/Mvar	Voltage Output/p.u.	Phase Angle/°
G_{A1}	762	166	1.03	0
G_{A2}	700	285	1.01	−17.0
G_{A3}	700	160	1.03	−12.9
G_{A4}	700	272	1.01	−25.9
G_{B1}	766	169	1.03	0
G_{B2}	700	292	1.01	−17.3
G_{B3}	700	154	1.03	1.8
G_{B4}	700	260	1.01	−11.5
L_{A7}	1100	100	0.96	−20.5
C_{A7}	0	0	0.96	−20.5
L_{A9}	1100	100	0.96	−29.6
C_{A9}	0	0	0.96	−29.6
L_{B7}	1700	100	0.96	−20.7
C_{B7}	0	0	0.96	−20.7
L_{B9}	1700	100	0.96	−15.7
C_{B9}	0	0	0.96	−15.7

5. Performances of the Two Strategies

5.1. Strategy 1—Relay Protection

It is assumed that a pole-to-ground fault occurs on the DC line between MMC 2 and MMC 4. At $t = 0$ s, the test system is in steady state. At $t = 10$ ms, a pole-to-ground fault occurs on the positive line side of B_{24}. Suppose the relay protection completes fault location detection at 20 ms. LCS of B_{24} and B_{42} starts action at the same time. UFD of B_{24} and B_{42} opens after 0.25 ms. At $t = 22.25$ ms, UFD achieves full contacts separation and MBs start action. During the entire process, if the arm current is higher than twice the rated IGBT current, the converter will be blocked. The rated IGBT current of MMC 1 and MMC 3 is 1.5 kA, and the rated IGBT current of MMC 2 and MMC 4 is 3.0 kA. The blocked MMCs deblock at $t = 40$ ms and operate according to the pre-fault control strategy. Figures 5–7 give the responses of B_{24}, B_{42}, and MMCs, respectively.

It can be seen from Figure 5 that for B_{24} (CB near the fault), the current when LCS starts action is 23.4 kA, and the current when MBs start action is 19.2 kA. After MBs are fully opened, the maximum voltage across them is 933.0 kV, which is 1.87 times the rated DC voltage. It can be seen from Figure 6 that for B_{42} (CB far from the fault), the current when LCS starts action is 5.5 kA. Furthermore, the current when MBs start action is 6.5 kA. The maximum voltage across MBs when they finish operation is 823.5 kV, which is 1.65 times the rated voltage of the DC grid. It can be seen from Figures 5 and 6 that it takes about 10 ms for the fault current to decay to zero.

It can be seen from Figure 7 that for MMCs near the fault, the currents of the smoothing reactors reach 14.3 kA and 7.3 kA, respectively. The arm currents are also very large, exceeding twice the rated IGBT current, and blocking MMC 2 and MMC 1. After MBs of B_{24} and B_{42} are disconnected, the overvoltage of the whole network reaches the peak value instantly. The voltage of MMC 3 reaches 1162.5 kV, which is more than twice the rated DC voltage (the peak value does not appear at the moment when CB is opened).

5.2. Strategy 2—Local Detection and Local Action

In this test system, the rated IGBT current of MMC 1 and MMC 3 is 1.5 kA, and the rated IGBT current of MMC 2 and MMC 4 is 3.0 kA. Therefore, MMC 1 and MMC 3 would be blocked when the arm current reaches 3.0 kA; MMC 2 and MMC 4 would be blocked when the arm current reaches 6.0 kA. The maximum currents of all DC lines in the test system under normal conditions are less than 3.0 kA. The defined positive directions of CBs are marked in Figure 4. Therefore, the HVDC CB is supposed to operate as soon as the current through it is in the positive direction and reaches 6.0 kA.

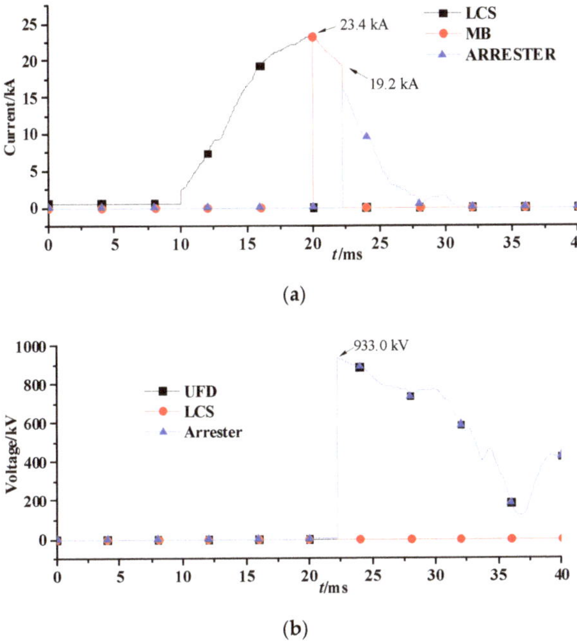

(a)

(b)

Figure 5. Responses of B_{24} under strategy 1: (**a**) currents flowing through load commutation switch (LCS), main breakers (MBs), and arrester; (**b**) voltages across ultra-fast disconnector (UFD), LCS, and arrester.

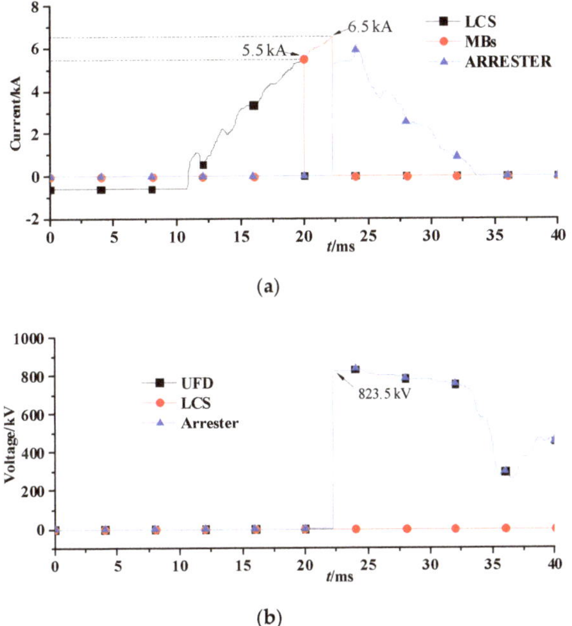

(a)

(b)

Figure 6. Responses of B_{42} under strategy 1: (**a**) currents flowing through LCS, MBs, and arrester; (**b**) voltages across UFD, LCS, and arrester.

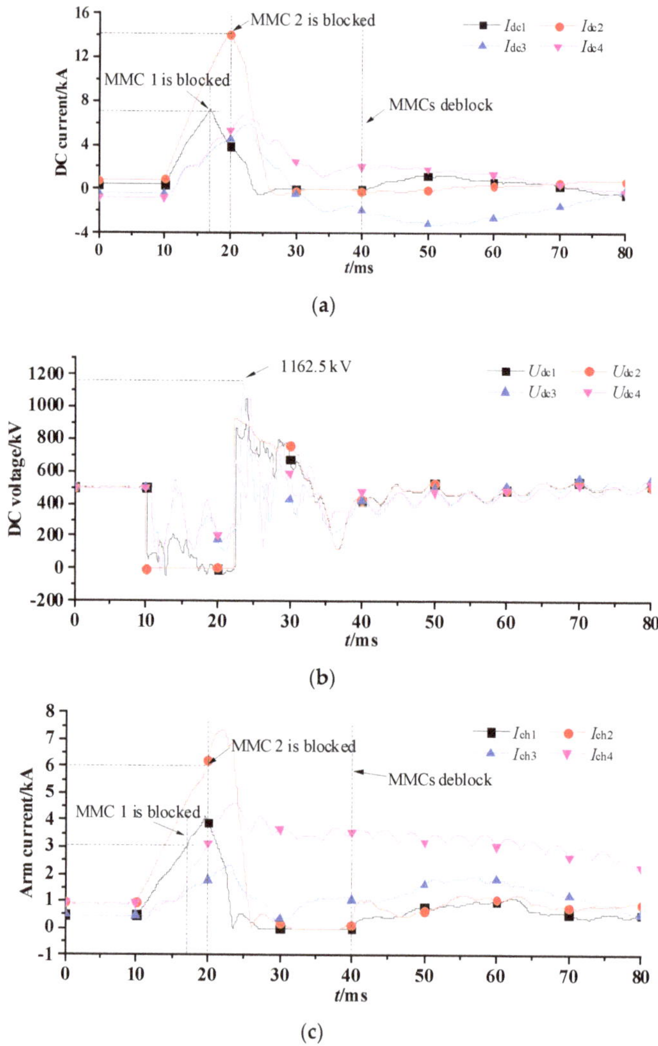

Figure 7. Responses of MMCs under strategy 1: (**a**) currents flowing through the smoothing reactors; (**b**) voltages across MMCs; (**c**) arm currents.

It is assumed that a pole-to-ground fault occurs on the DC line between MMC 2 and MMC 4. At $t = 0$ s, the test system is in steady state. At $t = 10$ ms, a pole-to-ground fault occurs on the positive line side of B_{24}. Figure 8 shows the currents flowing through the eight HVDC CBs. It can be seen that the current through B_{24} and B_{42} on both sides of the fault line reaches 6.0 kA. B_{24} and B_{42} start action at 1.6 ms and 8.8 ms after the fault, respectively. As a result, the DC currents passing through the other CBs begin to decrease, and the other CBs do not operate.

Then, the speed and selectivity of the proposed strategy is studied by complete line fault scanning of the test system. This is achieved by simulation of DC faults at all typical fault locations. The typical fault locations are selected as the sending end, the receiving end, and the midpoint of all the DC lines. The results are shown in Table 5.

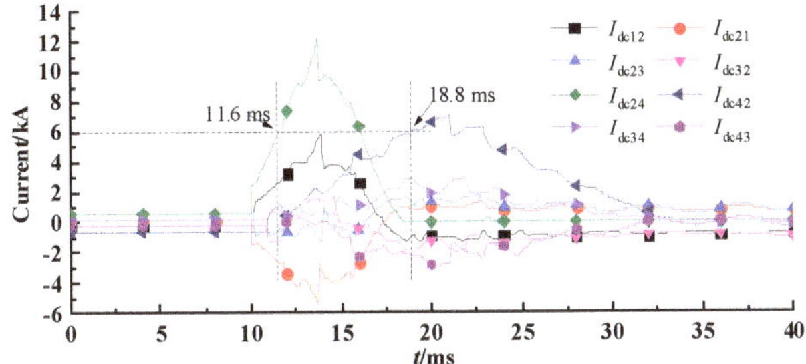

Figure 8. Currents flowing through eight HVDC breakers.

Table 5. The activated CBs and activated time under different faults.

Fault Location	Activated CBs and Activated Time (After Fault)/ms	
B_{13} line side	B_{13}: 1.6	B_{31}: 7.6
Midpoint of line 13	B_{13}: 3.8	B_{31}: 5.1
B_{31} line side	B_{13}: 6.3	B_{31}: 3.3
B_{12} line side	B_{12}: 2.5	B_{21}: 3.9
Midpoint of line 12	B_{12}: 3.6	B_{21}: 3.5
B_{21} line side	B_{12}: 3.9	B_{21}: 2.5
B_{24} line side	B_{24}: 1.6	B_{42}: 8.8
Midpoint of line 24	B_{24}: 4.1	B_{42}: 6.0
B_{42} line side	B_{24}: 6.9	B_{42}: 3.3
B_{34} line side	B_{34}: 2.3	B_{43}: 6.7
Midpoint of line 34	B_{34}: 4.9	B_{43}: 5.0
B_{43} line side	B_{34}: 6.8	B_{43}: 2.5

It is noted in Table 5 that this strategy has high speed and protection selectivity. Take the fault on the positive line side of B_{24} as an example for explanation. Obviously, CBs closest to the fault site are B_{24} and B_{21}. Therefore, B_{24} and B_{21} should be the first CBs to start action. However, B_{24} is activated but B_{21} is not. On the one hand, this is because the fault current through B_{21} is provided by MMC 1. Due to the smoothing reactors and DC line, the fault current through B_{21} reaches the operating current more than 2 ms later than B_{24}. Therefore, for B_{21}, it will not start action since the fault disappears once B_{24} starts action. On the other hand, the fault current through B_{21} is in the opposite direction, whereas the fault current through B_{24} is in the positive direction. Therefore, according to the local protection and local action strategy, B_{21} will not start action.

5.3. Performance Comparison of Two Strategies

To show the complete features of strategy 2, comparisons with strategy 1 are given under the pole-to-ground fault on the positive line side of B_{24}. Figures 9–11 give the responses of B_{24}, B_{42}, and MMCs, respectively. It can be seen from Figure 9 that for B_{24} near the fault, the current when LCS starts action is 6.0 kA. Moreover, the current when MBs start action is 12.6 kA. After MBs are fully opened, the maximum voltage across them is 890.6 kV, which is 1.78 times the rated voltage of the DC grid. It can be seen from Figure 10 that for B_{42} far from the fault, the current when LCS starts action is 6.0 kA. Furthermore, the current when MBs start action is 7.1 kA. The maximum voltage across MBs after they are disconnected is 831.9 kV, which is 1.66 times the rated DC voltage. Both Figures 9 and 10 show that it takes about 10 ms for the fault current to decay to zero.

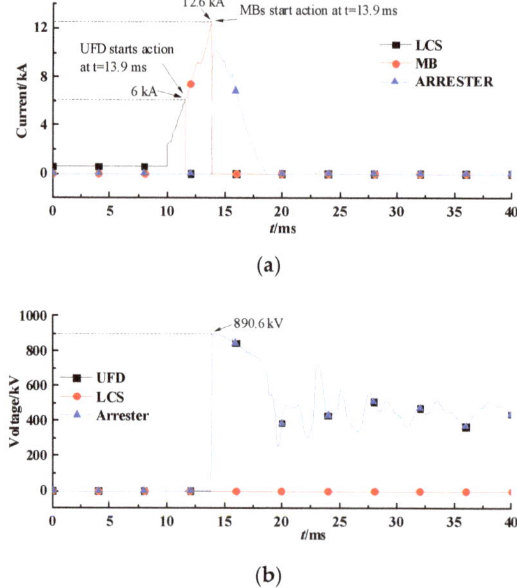

(**a**)

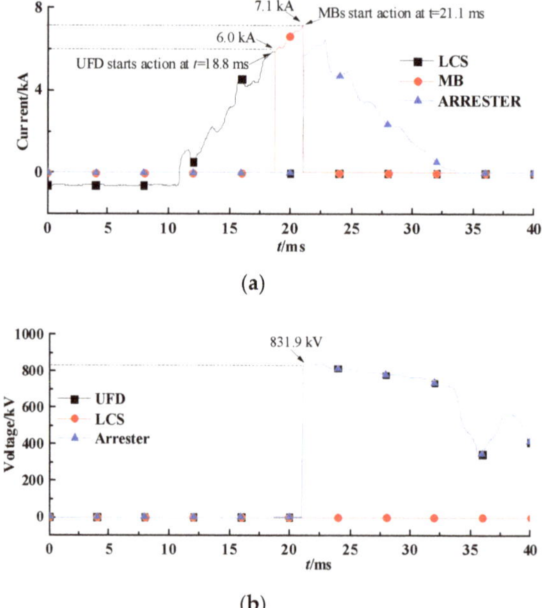

(**b**)

Figure 9. Responses of B_{24} under strategy 2: (**a**) currents flowing through LCS, MBs, and arrester; (**b**) voltages across UFD, LCS, and arrester.

(**a**)

(**b**)

Figure 10. Responses of B_{42} under strategy 2: (**a**) currents flowing through LCS, MBs, and arrester; (**b**) voltages across UFD, LCS, and arrester.

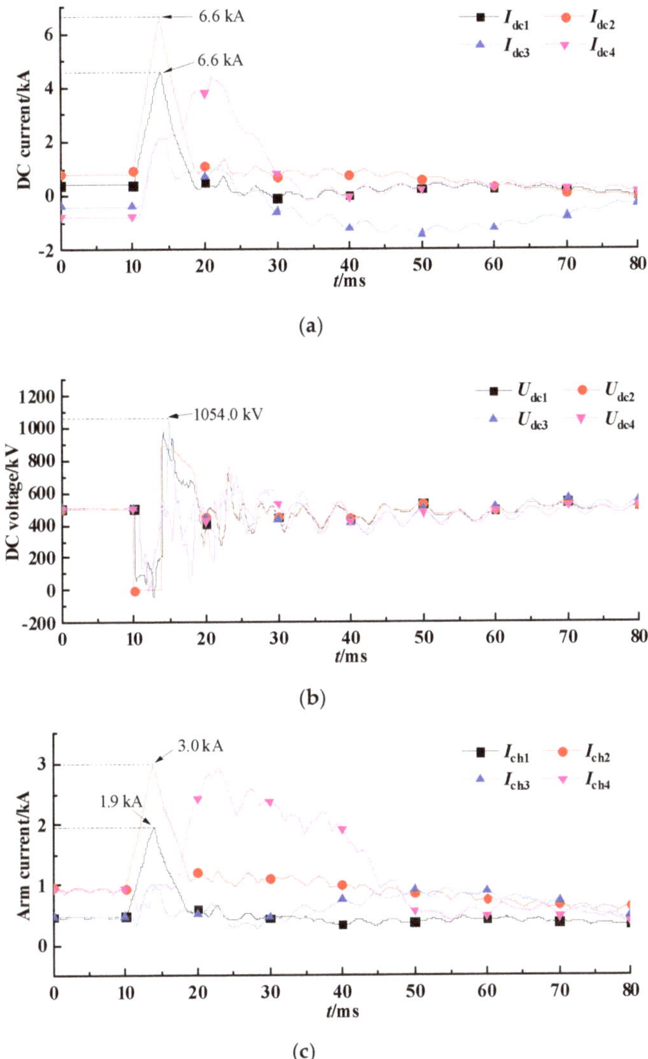

Figure 11. Responses of MMCs under strategy 2: (**a**) currents flowing through the smoothing reactors; (**b**) voltages across MMCs; (**c**) arm currents.

It can be seen from Figure 11 that for MMC 2 and MMC 1 that are close to the fault, the currents through the smoothing reactors reach 6.6 kA and 4.6 kA, respectively. The arm currents do not exceed twice the rated IGBT currents, and MMC 2 and MMC 1 do not need to be blocked. When B_{24} and B_{42} are completely disconnected, the overvoltage of the whole network reaches the peak value instantly. The voltage of MMC 3 reaches 1054.0 kV, which is more than twice the rated DC voltage (the peak value does not appear at the moment when CB is opened).

For the pole-to-ground fault on the positive line side of B_{24}, Table 6 shows the performances of the two strategies. It can be seen from Table 6 that the fault isolation time is earlier in strategy 2. The current level when CBs closest to the fault are activated is much lower in strategy 2 than that in strategy 1. In addition, in strategy 2, the peak currents flowing through the smoothing reactors and the arms are greatly reduced. More specifically, the MMCs in strategy 2 do not need to be blocked, and the impact of the fault is mitigated. Above all, the transient performance of strategy 2 is improved.

Table 6. Performance comparison of the two strategies.

Item	Strategy 1	Strategy 2
B_{24} LCS activated time after the fault/ms	10 ms	1.6 ms
The current level when B_{24} LCS is activated/kA	23.4 kA	6.0 kA
The current level when B_{24} MBs are activated/kA	19.2 kA	12.6 kA
The overvoltage times when B_{24} MBs interrupts the fault current	1.87	1.78
B_{42} LCS activated time after the fault/ms	10 ms	8.8 ms
The current level when B_{42} LCS is activated/kA	5.5 kA	6.0 kA
The current level when B_{24} MBs are activated/kA	6.5 kA	7.1 kA
The overvoltage times when B_{42} MB interrupts the fault current	1.65	1.66
Peak current through MMC 1 smoothing reactor/kA	7.3 kA	4.6 kA
Peak current through MMC 2 smoothing reactor/kA	14.3 kA	6.6 kA
Peak current through MMC 3 smoothing reactor/kA	6.0 kA	2.1 kA
Peak current through MMC 4 smoothing reactor /kA	7.0 kA	4.4 kA
Peak current through the 6 arms of MMC 1/kA	4.1 kA	1.9 kA
Peak current through the 6 arms of MMC 2/kA	7.4 kA	3.0 kA
Peak current through the 6 arms of MMC 3/kA	2.4 kA	1.0 kA
Peak current through the 6 arms of MMC 4/kA	4.6 kA	2.9 kA
The DC grid overvoltage times when B_{24} and B_{42} MBs interrupt the fault current	2.33	2.11

5.4. Cost Comparison of Two Strategies

Taking the pole-to-ground fault on the positive line side of B_{24}, for example, the device costs of CBs can be computed as follows. LCS is not required to withstand high voltages. MBs should be able to withstand the maximum pole-to-ground voltage as well as the maximum fault current after MBs are activated. Therefore, MBs are the main investment in both the two strategies [23].

Assume that the HiPak 5SNA 3000K452300 (ABB, Lenzburg, Switzerland) is adopted as the IGBT module. The rated collector–emitter voltage of the IGBT module is 4.5 kV. The breaking capability is 3 kA. Since the breaker is designed to break the current in either current direction, the direction index is 2. The peak current of the IGBT module can be obtained from Table 6. Moreover, the required breaking capability of CBs is usually larger than that. The IGBTs should be connected in parallel to guarantee sufficient current breaking capacity. In addition, series IGBTs in MBs need to withstand the pole-to-ground voltage. Furthermore, the series index could be calculated based on the maximum pole-to-ground voltage. In HVDC grids, one converter is usually connected to m ($m \geq 2$) DC lines. Therefore, the total required device number of B_{24} under the two strategies could be computed as listed in Table 7.

Table 7. Required device number of the two strategies.

	Strategy 1	Strategy 2
Current breaking ability	21 kA	15 kA
Main investment	MBs	MBs
Required number of CBs	m	m
Series index	226	215
Direction index	2	2
Parallel index	7	5
Number of required IGBTs	3164 m	2150 m

It can be seen from Table 7 that as the number of the connected DC lines rises, the number of required IGBTs in strategy 1 increases more. Moreover, it is the same for other CBs. The cost savings will be more conspicuous if there are more DC lines. Therefore, strategy 2 is very beneficial to reduce the investment of CBs in HVDC grids.

6. Conclusions

This paper proposes a local detection and local action strategy for DC grids. Compared with the conventional relay protection strategy, the proposed strategy has the following advantages:

(1) The proposed strategy has good protection selectivity. The difficulties in fault detection of DC grids are well solved;

(2) The proposed strategy has high protection speed, which can quickly isolate the faults, and greatly reduces the impacts of the faults on the DC system;

(3) The proposed strategy can greatly reduce the interrupted current level of HVDC CBs, and can greatly reduce the cost of HVDC CBs.

Author Contributions: Conceptualization, Z.Z. and Z.X.; methodology, Z.X.; software, Z.Z.; validation, J.Y.; formal analysis, J.Y.; investigation, J.Y.; resources, Z.X.; data curation, Z.X.; writing—original draft preparation, J.Y.; writing—review and editing, Z.Z.; visualization, Z.Z.; supervision, Z.X.; project administration, Z.X.; funding acquisition, Z.X. All authors have read and agreed to the published version of the manuscript.

Funding: This research was funded by the Headquarters Research Projects of State Grid Corporation of China, grant number SGJSJY00GHJS1900028.

Conflicts of Interest: The authors declare no conflict of interest.

References

1. Zhao, W.J. *HVDC Engineering Technology*, 2nd ed.; China Electric Power Press: Beijing, China, 2011.

2. Xu, Z.; Tu, Q.; Guan, M. *Voltage Source Converter Based HVDC Power Transmission Systems*, 2nd ed.; China Machine Press: Beijing, China, 2013.

3. Feldman, R.; Tomasini, M.; Amankwah, E.; Clare, J.; Wheeler, P.; Trainer, D.R.; Whitehouse, R.S. A Hybrid Modular Multilevel Voltage Source Converter for HVDC Power Transmission. *IEEE Trans. Ind. Appl.* **2013**, *49*, 1577–1588. [CrossRef]

4. Marquardt, R. Modular multilevel converter: A universal concept for HVDC-networks and extended DC-bus-applications. In Proceedings of the 2010 International Power Electronics Conference (IPEC), Sapporo, Japan, 21–24 June 2010.

5. Adam, G.P.; Ahmed, K.H.; Finney, S.J.; Bell, K.; Williams, B.W. New Breed of Network Fault-Tolerant Voltage-Source-Converter HVDC Transmission System. *IEEE Trans. Power Syst.* **2013**, *28*, 335–346. [CrossRef]

6. Ara, R.; Khan, U.A.; Bhatti, A.I.; Lee, B.W. A Reliable Protection Scheme for Fast DC Fault Clearance in a VSC-Based Meshed MTDC Grid. *IEEE Access* **2020**, *8*, 88188–88199. [CrossRef]

7. Lv, C.; Tai, N.; Zheng, X. Protection principle for MMC-HVDC DC lines based on boundary signals. *J. Eng.* **2017**, *13*, 961–965. [CrossRef]

8. Kenzelmann, S.; Rufer, A.; Dujic, D.; Canales, F.; de Novaes, Y. Isolated DC/DC Structure Based on Modular Multilevel Converter. *IEEE Trans. Power Electron.* **2015**, *30*, 89–98. [CrossRef]

9. Suo, Z.; Li, G.; Xu, L.; Li, R.; Wang, W.; Chi, Y. Hybrid modular multilevel converter based multi-terminal DC/DC converter with minimised full-bridge submodules ratio considering DC fault isolation. *IET Renew. Power Gener.* **2016**, *10*, 1587–1596. [CrossRef]

10. Wen, J.; Tang, Y.; Xie, C.; Chen, X.; Wang, Y.; An, N. Research on Power Flow Algorithm of Power System with UPFC. In Proceedings of the 2018 International Conference on Power System Technology (POWERCON), Guangzhou, China, 6–8 November 2018.

11. Rouzbehi, K.; Candela, J.; Luna, A.; Gharehpetian, G.; Rodriguez, P. Flexible Control of Power Flow in Multiterminal DC Grids Using DC–DC Converter. *IEEE J. Emerg. Sel. Top. Power Electron.* **2016**, *4*, 1135–1144. [CrossRef]

12. Wu, Y.; Lv, Z.; He, Z. Study on the protection strategies of HVDC grid for overhead line application. *Proc. CSEE* **2016**, *36*, 3726–3734.

13. Wei, X.; Gao, C.; Luo, X.; Zhou, W.; Wu, Y. A novel design of high-voltage DC circuit breaker in HVDC flexible transmission grid. *Autom. Electr. Power Syst.* **2013**, *37*, 95–102.

14. Li, B.; He, J. Research on the DC fault isolating technique in multi-terminal DC system. *Proc. CSEE* **2016**, *36*, 87–95.

15. CIGRE B4-52 Working Group. *HVDC Grid Feasibility Study*; International Council on Large Electric Systems: Mel-Bourne, Australia, 2011.

16. Callavik, M.; Blomberg, A.; Hafner, J. *The Hybrid HVDC Breaker-An Innovation Breakthrough Enabling Reliable HVDC Grids*; ABB Grid Systems: Zurich, Switzerland, 2012; pp. 1–10.

17. Hafner, J.; Jacobason, B. Proactive hybrid HVDC breakers-a key innovation for reliable HVDC grids. In Proceedings of the CIGRE 2011 Bologna Symposium, Bologna, Italy, 13–15 September 2011.

18. Shukla, A.; Demetriades, G. A Survey on Hybrid Circuit-Breaker Topologies. *IEEE Trans. Power Deliv.* **2015**, *30*, 627–641.

19. Li, L.; Zhang, B.; Fang, C.; Li, W.; Ren, X.; Diao, T. Researches on interruption characteristics of a hybrid HVDC circuit breaker. In Proceedings of the 2017 4th International Conference on Electric Power Equipment-Switching Technology (ICEPE-ST), Xi'an, China, 22–25 October 2017.

20. Lazzari, R.; Piegari, L. Design and Implementation of HVDC Hybrid Circuit Breaker. *IEEE Trans. Power Electron.* **2019**, *34*, 7369–7380. [CrossRef]

21. Song, G.; Gao, S.; Cai, X.; Zhang, J.; Rao, J.; Suonan, J. Survey of relay protection technology for HVDC transmission lines. *Autom. Electr. Power Syst.* **2012**, *36*, 123–129.

22. Kundur, P. *Power System Stability and Control*; McGraw-Hill Incorporated: New York, NY, USA, 1994.

23. Liu, G.; Xu, F.; Xu, Z.; Zhang, Z.; Tang, G. Assembly HVDC Breaker for HVDC Grids with Modular Multilevel Converters. *IEEE Trans. Power Electron.* **2017**, *32*, 931–941. [CrossRef]

Article

An Adaptive Protection for Radial AC Microgrid Using IEC 61850 Communication Standard: Algorithm Proposal Using Offline Simulations

Aushiq Ali Memon * and Kimmo Kauhaniemi

School of Technology and Innovations, University of Vaasa, Wolffintie 34, FI-65200 Vaasa, Finland;
Kimmo.Kauhaniemi@univaasa.fi
* Correspondence: aushiq.memon@univaasa.fi or aushiq_37@yahoo.com; Tel.: +358-414-744-093

Received: 23 August 2020; Accepted: 9 October 2020; Published: 13 October 2020

Abstract: The IEC 61850 communication standard is getting popular for application in electric power substation automation. This paper focuses on the potential application of the IEC 61850 generic object-oriented substation event (GOOSE) protocol in the AC microgrid for adaptive protection. The focus of the paper is to utilize the existing low-voltage ride through characteristic of distributed generators (DGs) with a reactive power supply during faults and communication between intelligent electronic devices (IEDs) at different locations for adaptive overcurrent protection. The adaptive overcurrent IEDs detect the faults with two different preplanned settings groups: lower settings for the islanded mode and higher settings for the grid-connected mode considering limited fault contributions from the converter-based DGs. Setting groups are changed to lower values quickly using the circuit breaker status signal (XCBR) after loss-of-mains, loss-of-DG or islanding is detected. The methods of fault detection and isolation for two different kinds of communication-based IEDs (adaptive/nonadaptive) are explained for three-phase faults at two different locations. The communication-based IEDs take decisions in a decentralized manner, using information about the circuit breaker status, fault detection and current magnitude comparison signals obtained from other IEDs. However, the developed algorithm can also be implemented with the centralized system. An adaptive overcurrent protection algorithm was evaluated with PSCAD (Power Systems Computer Aided Design) simulations, and results were found to be effective for the considered fault cases.

Keywords: AC microgrid; adaptive protection; IEC 61850 GOOSE protocol; substation automation

---

1. Introduction

According to the CIGRE C6.22 working group definition, microgrids are electrical distribution systems containing loads and distributed energy resources (DERs) like distributed generators (DGs) (renewable/nonrenewable), energy storage devices or controlled loads that can be operated in a controlled and coordinated way either while connected to the main power network or while islanded [1]. Microgrids can be classified as either AC microgrids, DC microgrids or AC/DC hybrid microgrids, each having their own advantages, limitations and challenges, as described in [2]. The technical challenges of AC microgrids can be broadly divided into two main categories: control challenges and protection challenges. The protection challenges can be further divided into two categories according to operational modes of the AC microgrid: grid-connected mode and islanded mode protection challenges.

When the AC microgrid is operated in the grid-connected mode, a large magnitude of fault current (ten times the full-load current or more) is available from the main grid in order to activate the overcurrent protection devices within the AC microgrid. When the AC microgrid is operated in the islanded mode, a very low magnitude of fault current is available from DGs within the AC microgrid, and hence, overcurrent devices with a single setting become insensitive. The consequences are the miscoordination of overcurrent devices, longer tripping delays and even no trips at all during different fault situations. The magnitude and duration of the fault current is mainly limited by the control of the converter-based DGs within the AC microgrid, which can be overcome by an additional fault-current source (FCS), like an energy storage device with high short-circuit capacity, and thus, single-setting overcurrent devices will become effective. However, the connection of an additional FCS will make the protection scheme unreliable due to dependence upon the single energy storage device. Moreover, the connection of many such FCSs will make the scheme quite expensive [3]. Another alternative approach for using only single-setting overcurrent devices can be the limitation of the fault current from the main grid or directly coupled DGs using fault-current limiters (FCLs) in grid-connected mode and using lower fault-current trip settings, which can also work in islanded mode with reduced short-circuit currents. This approach causes overcurrent devices to be more sensitive in grid-connected mode and prone to nuisance tripping during transient events [4]. The huge difference of the fault-current magnitude and duration in grid-connected and islanded mode calls for adaptive protection schemes for the AC microgrid.

The adaptive protection schemes can be only overcurrent-based [5] or a combination of overcurrent-based and unit type (current differential) or based on other new protection methods like traveling waves-based [3]. The adaptive overcurrent protection necessarily requires such overcurrent devices that provide the flexibility for changing the tripping settings like numerical overcurrent (OC) relays with several setting groups [5]. The overcurrent schemes can be used more effectively in AC microgrids with the majority of directly coupled DGs (synchronous generators) compared with only the converter-based DGs, since the latter provide very limited fault currents for a very limited duration of time. Another reasonable adaptive approach is to use only the overcurrent protection scheme in the grid-connected mode and other protection schemes like directional overcurrent, harmonic content-based, voltage-based, symmetrical component-based, etc. in islanded mode for the AC microgrid with the converter-based DGs, with all functions included in a single protection device called the IED (intelligent electronic device). However, the protection schemes proposed for the islanded mode are not effective in every fault situation, and the majority of them need high-speed communication to remain effective [4]. Finding a suitable and cost-effective combination of different effective protection schemes for the islanded mode with the converter-based DGs to work as primary and backup protection in a coordinated manner is still a huge challenge. An adaptive protection can be implemented either in a centralized manner by using a microgrid central controller to change the active-group settings [5] or in a decentralized manner in which IEDs in the microgrid change their own active-settings groups by receiving a trip-signal/breaker status from another IED or circuit breaker. The centralized adaptive protection scheme necessarily requires a redundant microgrid central controller to maintain a certain level of reliability. For a decentralized adaptive protection scheme, the IEDs must be equipped with the required intelligent agents and logics in order to perform various functions in an autonomous manner using the available information (data/measurements/signals) both locally and remotely.

Previously, the adaptive protection for the AC microgrid using centralized protection and communication architecture was proposed in [5–7]. An adaptive overcurrent protection for microgrids using inverse-time directional overcurrent relays (DOCRs) was presented in [8]. In this paper, artificial neural networks (ANNs) at the central human machine interface (HMI) or data concentrator are implemented for the detection and location of the faults. The protection coordination of OC relays using the linear programming approach is presented for the radial and looped configurations of microgrids in both the grid-connected and islanded modes. An adaptive protection combined with machine learning for medium voltage (MV) microgrids was reported in [9]. The proposed methodology requires a database available beforehand, which has been obtained through simulation in this research. Then, using the data mining methodology, the meaningful information is extracted quantitatively from the database. The ANN is used for fault detection and support vector machine (SVM) for fault location. The proposed method also requires relay settings calculations and recordings in the control center or relays beforehand. Moreover, the proposed scheme may generate inaccuracies in the case of data corruption, and therefore, additional countermeasures will be required. A new adaptive protection coordination scheme based on the Kohonen map or self-organizing map (SOM) clustering algorithm was proposed recently in [10] for the inverse-time OC relays. In this paper, the protection coordination is improved gradually in the three phases of the proposed algorithm, namely conventional, clustering and sub-clustering phases. The proposed method uses digital OC relays with four setting groups. The performance of the method was presented in terms of the total miscoordination time (TMT) index using a modified IEEE 33-bus network with two synchronous generators and two electric vehicle (EV) charging stations. A decentralized adaptive protection scheme using DOCRs, teleprotection and a fuzzy system in real time was proposed in [11] for the transmission system. In this paper, the transient stability constraint satisfying the maximum operating time of DOCRs was considered. Due to the dynamic adaption of the fuzzy system to the changing system conditions, the actuation time of relays was decreased, keeping the stability and coordination intact. An optimal overcurrent relay coordination in the presence of inverter-based wind farms and electrical energy storage devices was presented in [12]. In this paper, the optimal protection coordination of inverse-time DOCRs with varying load demands and changing unit commitments of DGs is presented using mixed integer nonlinear programming. A hybrid particle swarm optimization-integer linear programming (PSO-ILP) algorithm was suggested recently in [13] for the proper coordination of OC relays by suggesting proper settings groups for the changing network states. The adaptive differential protections for wind farm-integrated networks were reported in [14,15]. However, the differential protection inherently cannot provide the backup protection, and usually, the time-coordinated overcurrent protection is used as the backup protection.

The modeling of the inter-substation communication based on the IEC 61850 standard was presented in [16] for the differential protection (Sampled Values (SV) messages) and in [17] for the distance protection (generic object-oriented substation event (GOOSE) messages). In both [16] and [17], the virtual simulated communication networks were used based on a non-real-time tool called the riverbed modeler network software. In both references [16] and [17], the tunneling communication mechanism between substations was used for the differential and the distance protection functions, respectively. In [16], it was evaluated that the dedicated fiber optic network link had better performance in terms of the end-to-end delay of SV and GOOSE messages compared with an asynchronous transfer mode (ATM) link and synchronous optical networking (SONET) links. It was concluded in [17] that the links with lower bandwidths were not suitable for long distances; however, a more accelerated distance protection can be implemented, even with lower bandwidth links, compared with the conventional distance protection scheme. An adaptive protection system based on the IEC 61850 for MV smart grids was presented in [18]. In this paper, the dynamic publisher/subscriber reconfiguration of the protection devices for the implementation of the advanced fault location, isolation and service restoration (FLISR) was suggested. Since, the remote changes of the IED settings are not supported by the current versions of the IEC 61850 standard, therefore, the change of the operational settings after the network reconfiguration was suggested using the exchange of MMS (manufacturing message specification) messages with IEDs. Additionally, the logic selectivity was proposed to support remote changes of GOOSE communication schema without interrupting the FLISR operation. A mixed-layer 2/3 approach was also suggested in the paper to support both the MMS and the GOOSE implementations for the field demo of an Italian MV network. A detailed survey of different adaptive protections of microgrids was presented recently in [19]. For a further detailed review of different microgrid protection schemes, their challenges and developments, the recent review articles [20–23] are suggested, in addition to the previous review article [4] by the authors. For further information related to IEC 61850-based substation automation systems and related issues, the recent literature survey done in [24] is also recommended.

Based on the recent literature review presented above, it was found that less literature is available for the role of IEC 61850 standard-based communication in the protection coordination of the AC microgrids with decentralized protection and communication architecture. Moreover, a low-voltage ride through (LVRT) capability with reactive power support from the converter-based DGs in the case of AC microgrid faults has rarely been used for adaptive protection. The high risks of communication link failures and unacceptable and unpredictable communication delays are still the limiting factors to use communication links for high-speed protection functions. However, the use of a communication link is inevitable for protection functions like transfer trips from the breaker/IED at the point of common coupling (PCC) to another breaker/IED within the AC microgrid for loss of mains detection and changing preplanned active-groups settings/functions during the transition from the grid-connected to island mode for deactivating sensitive anti-islanding protection during faults and for reverse interlocking schemes. In this paper, the main focus is to discuss how an IEC 61850 communication can be applied for a decentralized preplanned adaptive overcurrent protection in a radial AC microgrid. Additionally, the DGs with LVRT capability and reactive power support in islanded mode are considered in order to implement the adaptive overcurrent protection.

The rest of the paper is organized in a manner that Section 2 presents adaptive protection based on the IEC 61850 communication standard by explaining a generalized architecture of the adaptive AC microgrid. Section 3 gives a case study background of the adaptive protection of a radial AC microgrid, explaining GOOSE (generic object-oriented substation event) message delays (transfer time) for IED to IED communication for different functions, the schematic diagram of radial the AC microgrid and adaptive protection settings of different IEDs. Section 4 explains the details of the proposed adaptive protection methods and results for both the grid-connected and islanded modes of operation. Additionally, the control of DGs and the LVRT capability of DGs are also explained in this section. Section 5 gives a brief discussion about the previous methods, the contribution of the research presented in this paper and what is needed for the practical implementation of the proposed method in the future. Section 6 provides the conclusion of the paper.

2. Adaptive Protection Based on IEC 61850 Communication Standard

An adaptive protection is necessarily required for AC microgrids due to changing operational modes (grid-connected and islanded), due to the formation of controlled islands due to faults within the AC microgrid, due to intermittent DGs and periodic load variations and due to the economical operations of the AC microgrid [4,25]. An adaptive protection is defined as an online activity that changes to the preferable protection device response for modified system conditions or requirements. An adaptive protection is normally automated, but some timely human interventions may also be included. Adaptive relay is a relay that includes various settings, characteristics or logic functions capable of speedy online modifications by means of externally generated signals or control actions [26]. The modern intelligent electronic devices (IEDs) not only provide various protection functions (overcurrent, over/under voltage, etc.) integrated in a single physical device but, also, offer various setting groups for each of the available protection functions. The various setting groups of the protection functions can be modified or altered in an adaptive manner using the communication link between IEDs and IEDs and circuit breakers (CBs). Recently, the popularity of the IEC 61850 communication standard for application in electric power substation automation has increased considerably due to its promise of interoperations among IEDs from different manufacturers. The initial focus of the standard is on communication between IEDs within a single substation, but its extension for communication between several substations in the future is possible. The IEC 61850 standard explains the standardized structures for the data model and definitions of rules for the exchange of these data. IEDs from different manufacturers that comply with these standard data model definitions can then communicate, understand and interact with each other [26]. The IEC 61850 standard as a common protocol enables the integration of all protection, control, measurement and monitoring functions [27].

The generalized architecture for adaptive AC microgrid protection based on the IEC 61850 communication standard is depicted in Figure 1. The IEC 61850 communication architecture for adaptive AC microgrid protection can be subdivided into three levels: process level, bay level and substation level. At the process level, the electrical parameters measurement data (MMXU) from the voltage and current sensors (VTs and CTs) and status of the circuit breakers (XCBR) inside the AC microgrid will be collected and digitized by merging units (MUs). At the bay level, the IEDs for lines, DGs and loads of the AC microgrid will collect the digitized measurement data (MMXU) and circuit breaker status signals (XCBR) from the process bus. Each MU will publish data to process the bus, and each IED will subscribe to the respective published data from the process bus. Each of the line, DG and load IED will use measurement data (MMXU) from their respective MU for performing the active protection function like overcurrent protection in the case of faults. The status signal of the circuit breakers inside the AC microgrid (XCBR) will be used by each adaptive IED to change the active setting groups of the protection function in the case of a fault inside the AC microgrid in islanded mode. Moreover, a XCBR signal can also be used for the transfer trip of nonadaptive IEDs that are unable to detect the faults within the islanded AC microgrid. All IEDs at the bay level will also receive the status signal (XCBR_pcc) from the circuit breaker at the point of common coupling through the station bus at the substation level. The status signal from the PCC breaker (XCBR_pcc) will be used by each adaptive IED within the AC microgrid to change the active setting group of the protection function from grid-connected mode settings to islanded-mode settings and vice versa. The signal (XCBR_pcc) can also be used for the detection of the loss-of-mains event by DG IEDs and to deactivate the sensitive loss-of-mains protection functions in order to maintain stability and reliability of supply within the AC microgrid during the transition from the grid-connected to islanded mode. The station bus at the substation level will provide primary communication between the various logical nodes of IEDs. In other words, all IEDs at the bay level will communicate and share data/information (MMXU, XCBR, and XCBR_pcc) with each other using the station bus. At the station bus, a remote access point will also exist to share data with remote clients (for wide-area measurement and protection, etc.) through a wide-area network (WAN). This remote access point will implement security functions like data encryption and authentication for all data transfers and, thus, will unburden the individual IEDs to perform these tasks.

For an adaptive OC protection, the coordination between the control and protection of the AC microgrid will also be required, and control action will be required first, followed by protection action. In the grid-connected mode, a high fault current from the grid will be available, so depending on the protection settings of IEDs, it may be required to limit the magnitude of the fault current by the activation of FCLs, and in the islanded-mode with converter-based DGs, the enhancement of the fault current magnitude may be required by the activation of additional FCSs. The numerical results presented in [28] indicate that a majority of the photovoltaic (PV) inverters contribute a fault current of 200% or less for a duration of only an initial half-cycle and 110% of the rated current for an additional duration of 10 cycles or less. It is mentioned in [29] that the grid-connected converters can feed fault currents of 1.1–1.5 p.u. of their nominal currents. It should here be noted that extra FCSs like batteries, flywheels or supercapacitors with quick response times ($\leq$10 ms) [30] will either be necessarily required to support some type of DGs like photovoltaic DGs for providing standard LVRT capability or extending the LVRT duration of other types of DGs like wind turbine generators (WTGs) for proper protection coordination if the WTG is not capable of providing LVRT. The results presented in [31] show that a wind turbine of 1 MW can provide a fault current of magnitude equal to 120% of the rated current for seven cycles of supply frequency. This duration of seven cycles with a 50-Hz supply frequency is approximately equal to the initial duration of 150 ms after fault in the LVRT characteristic of the German BDEW-2008 standard [32]. Although the duration of 150 ms looks sufficient for the maintenance of proper protection coordination between two successive IEDs within the AC microgrid, assuming high speed communication with 3–10-ms one-way fast trip message transfer as per the IEC 61850 standard and high-speed circuit breakers (one-cycle operation). However, in some cases like data loss in the transmission channel, the retransmission of the message is required, which will result in an additional delay. Moreover, the coordination between various IEDs for breaker failure protection may be required. In such situations, the extension of the initial duration after the fault in the LVRT curve beyond 150 ms will be required, and hence, additional FCSs (flywheels or supercapacitors) will be required. In addition to that, a redundant communication and redundant synchronization clock architecture will be required to cover the communication link and synchronization clock failures as recommended in [33].

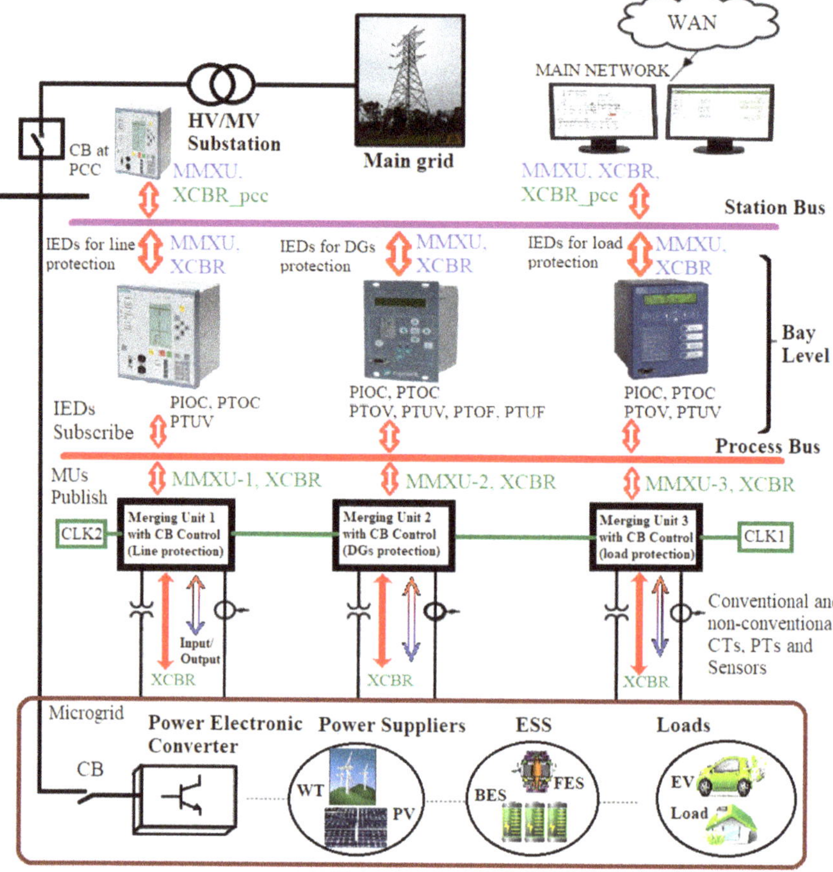

Figure 1. Adaptive AC microgrid protection based on the IEC 61850 communication standard. HV: High Voltage, WT: Wind Turbine, ESS: Energy Storage System, BES: Battery Energy Storage, FES: Flywheel Energy Storage, MMXU: Measurement, XCBR: Circuit Breaker, CT: Current Transformer, PT: Potential Transformer (VT: Voltage Transformer), CLK: Clock, PIOC: Instantaneous Overcurrent Protection, PTOC: AC Time Overcurrent Protection, PTOV: (Time) Overvoltage Protection, PTUV: (Time) Undervoltage Protection, PTOF: Overfrequency Protection, PTUF: Underfrequency Protection.

In this paper, the main focus was given to the adaptive OC protection using fault contributions from DGs with LVRT capability, particularly in the islanded mode of the AC microgrid. Hence, the control of DGs is not discussed in detail, except a few control actions for maintaining the voltage and frequency at the islanded sections, as explained in Section 4. Moreover, the loads and generation are considered balanced in islanded mode of the AC microgrid. The same is true even for the islanded MV and LV (low voltage) sections of the AC microgrid. The paper is limited to single fault events (three-phase short-circuit faults only) during the grid-connected and islanded modes with smooth transitions to islands. However, the method presented can be extended to other types of faults. In this paper, it is not considered how the islanded AC microgrid is reconnected back to the main grid after the removal of the fault events, which is mainly related to resynchronization procedures and not directly related to AC microgrid protection. Considering the previous research, the fault contributions from DGs (both PV and WTG) are taken as 1.2 p.u. or 120% of their rated nominal currents for a duration of 150 ms after the fault. During the initial fault duration of 150 ms, the active, passive and other islanding detection and protection schemes are considered normally interlocked and can be activated quickly after the loss of communication. This means that the anti-islanding protections like under-voltage protection at DG locations can be set by default to only detect fault conditions but not trip, and DGs start providing fault currents instantaneously according to the LVRT characteristics. The trip-blocking signal to anti-islanding protection can be sent additionally from an IED at PCC after a fault is detected on the main grid side; it should be done as fast as possible and within 3 ms after fault detection, as per the IEC 61850 standard. In this paper, the term "adaptive IED" mainly refers to the communication-assisted definite time overcurrent (DTOC) relay with two preplanned setting groups: higher setting group for the grid-connected mode and lower setting group for the islanded mode of operation. The case study of a typical radial AC microgrid equipped with adaptive DTOC relays and DGs with LVRT capability is presented in the next section.

3. Case Study

An IED-to-IED GOOSE message exchange within a substation is required for fast bus tripping in the case of bus faults and the interlocking of bus-IED in the case of feeder/line faults, the protection scheme traditionally known as the reverse interlocking scheme. The IED-to-IED GOOSE messages can also be used in the case of breaker failure to trip the adjacent breaker(s). This can be done by sending a trip command message to adjacent breakers from a protection IED with a built-in breaker failure function or from a dedicated IED performing only the breaker failure function. The transfer trip may also be required between two substations. The transfer time requirement of 10 ms was set in IEC 61850 for fast trip messages (releases and status changes) between substations (transfer time class TT5) and 3 ms for fast trips and blocking messages between IEDs within a substation (transfer time class TT6) [33,34]. The transfer time requirement also varies with respect to the specific protection function. The transfer times required for various protection functions are given Table 1.

Although very strict time requirements have been demanded in IEC 61850 for type 1A fast trip messages, in this study, an average transfer time of 10–20 ms was considered for the one-way GOOSE message to cover the limitations (the limited failures of LAN within a maximum allowed transfer time of 18 ms), safety margins (errors in the time-stamp accuracy) and redundant GOOSE messages for communication between substations, as explained in [35]. IEC 61850-90-1 [36] recommends a maximum time delay of 5 to 10 ms on the communication channel depending on the voltage level [37]. However, in order to meet the requirements of security, reliability and dependability according to the IEC 60834 standard, the communication system should meet the 3-ms transfer time requirement for 99.9999 percent of the time and should have a delay no longer than 18 ms for the remainder of the time [38]. A fixed transfer time of 20 ms is thus used for both IED-to-IED communication within a substation and IED-to-IED communication between different substations in this study to cover even the worst time delay of 18 ms for the type 1A messages. In practical cases, generally, the transfer time for communication between IEDs at different substations is longer than the transfer time within the same substation. Measuring the one-way communication latency by a round-trip time between two remote substations was discussed in [39]. The selected one-way transfer time of 20 ms for producing results corresponded to the fast messages of type 1B (the ideal case) with performance class P2/P3 (transfer time class TT4) [33,35], and it covers the worst-case delay of 18 ms of type 1A fast messages according to the IEC 60834 mentioned above. The considered transfer time was also within the range of practical observed time delays in the light-weight implementation of the IEC 61850 standard-based GOOSE messages done in [40]. Although GOOSE messages apply the heartbeat messages and an IED will issue a so-called burst of GOOSE messages right after the detection of the fault, for the final trip decision, the IED necessarily needs to know the updated status at the downstream IED after the fault to ensure proper coordination. The selected 20-ms communication delay between IEDs ensures that even the type 1A GOOSE messages with the maximum delay of 18 ms are also subscribed by IED for the trip decision. Another reason of selecting a 20-ms delay between IEDs is the potential requirement to use GOOSE messages to transfer analog data between IEDs for trip decision criterion like vectors of measured values (RMS values), which need to be transmitted only once per cycle of 50-Hz frequency. It means a new analog measurement data is required to be transmitted in just every 20 ms [37]. In the earlier publication [33], it was mentioned that there were two types of GSE (generic substation event) messages: GSSE (generic substation state event) message and GOOSE message. The GSSE message is the old binary-only message type. All the modern systems use the more flexible GOOSE message, which transfers both binary and analog data. Both GSSE and GOOSE can coexist but are not compatible with each other. The proposed protection algorithm in this paper not only uses the binary data but, also, uses the analog data (RMS magnitude of currents) for the trip decision.

Table 1. Protection functions, logical nodes and performance requirements as per IEC 61850-5:2003 (Annex G) [34].

Function	Performance Transfer Time (ms)	Corresponding LNs (Decomposition)	Starting Criteria/Remarks
Distance protection (PDIS)	5–20	IHMI, ITCI, ITMI, PDIS, TCTR, TVTR, XCBR, other primary equipment-related LNs	The monitoring part of the function is set into operation if the function is started. The function issues a start (pickup) signal in the case of an alert situation (impedance crosses boundary 1) and a trip in case of an emergency situation (impedance crosses boundary 2).
Bay interlocking	10	IHMI, ITCI, CILO, CSWI, XCBR, XSWI, (PTUV)—if applicable	The recalculation of interlocking conditions starts by any position change of the switchgear (circuit breaker, isolator, and grounding switch). Depending on the implementation, the recalculation may start not before switchgear selection.
Station-wide interlocking	-Blocking and release: 10 -Reservation: 100	IHMI, ITCI, CILO, CSWI, XCBR, XSWI, (PTUV)—if applicable	Position change of a switching device or request of the command function.
Breaker failure	5 (Delay settable ≤ 100)	IHMI, ITCI, ITMI, P … , RBRF, TCTR, CSWI	If a breaker gets a trip signal by some protection (for example, line protection) but does not open because of an internal failure, the fault has to be cleared by the adjacent breakers. The adjacent breakers may include breakers at remote substations (remote line ends). For this purpose, the breaker failure protection is started by the protection trip and supervises if the fault current disappears or not. If not, a trip signal is sent to all adjacent breakers after a preset delay.
Automatic protection adaptation (generic)	1–100 (Depending on the considered function)	IHMI, ITCI, ITMI, P …	-The protection trip makes the breaker failure protection alert, and the fault is cleared by adjacent breakers. The protection specialist may change the protection parameters (settings) if needed by static or slow predictable power system reconfiguration. -If the conditions for protection change dynamically during operation, the parameters of the protection may be changed by local or remote functions. Very often, complete pretested sets of parameters are changed rather than single parameters. -Changes in conditions are detected and communicated by some other functions, and the protection function is adapted to the changed power system condition.
Reverse blocking function (OC relays)	5	IHMI, ITCI, ITMI, P … (more than one)	- When a protection is triggered by OC: it sends blocking signal to upstream protections. it trips/opens its associated breaker if it does not receive a blocking signal issued by downstream protection.

In this paper, the conventional GOOSE (tunneled-GOOSE) messages in layer 2 (horizontal communication) with an Ethernet link are considered, because the short distances (a few km) between substations are considered. However, for the longer distances between substations where an Ethernet link is not possible, the routable-GOOSE (R-GOOSE) messages in layer 3 (vertical communication) for wide-area or system protection applications could be used, especially with wireless communication technologies using synchrophasors in compliance with IEC TR 61850-90-5. Some applications of R-GOOSE were reported in [41]. The normal predefined fixed GOOSE message transfer delay of 40 ms (2 cycles of 50-Hz power system) was assumed previously for the adaptive protection of the microgrid using communication over a WiMAX network in [42], and the actual latency observed was within 35 ms with no data packet loss. However, due to packet loss and, consequently, retransmissions, the overall delay could further increase, thus limiting the application of WiMAX (R-GOOSE) to comparatively slower control and protection actions like status updates and protection settings during scheduled maintenance and load management. The adaptive protection methodology presented in this paper is concerned with primary and backup protections of the microgrid during faults in predefined operational modes: grid-connected or islanded modes. With this regard, a communication-dependent coordination methodology is proposed in Section 4 for the cases when the fault happens between two defined IEDs in grid-connected and islanded modes. The proposed methodology is very generic in nature and can be implemented in any protection IED.

The schematic diagram of a radial AC microgrid under study is shown in Figure 2. The considered AC microgrid consists of one MV bus of 20 kV (Substation Bus-2) and one LV bus of 0.4 kV (Substation Bus-3). A load of 2 MW at Substation Bus-2 is supplied by a wind turbine generator (WTG) of 2-MVA capacity, whereas a load of 0.4 MW at Substation Bus-3 is supplied by a photovoltaic (PV) generator of 0.4 MVA. The MV bus (Substation Bus-2) of the AC microgrid is connected with the LV bus (Substation Bus-3) of the AC microgrid through a 1-km-long, 20-kV cable line and 0.5-MVA, 20/0.4-kV transformer. The AC microgrid is connected with the main grid through a 2-km-long, 20-kV overhead line and an intermediate 20-kV Substation Bus-1. The WTG is connected to Substation Bus-2 through a 0.2-km-long 20-kV cable and a 2-MVA, 0.69/20-kV transformer (inside the WTG model). A 2-km overhead line between Substation Bus-1 and Substation Bus-2 is protected by two circuit breakers, CB1 and CB2, with the related protection IEDs. The protection IED1 is considered to be a nonadaptive IED due to its direct connection with the main grid, whereas the protection IED2 is considered as an adaptive IED. In the grid-connected mode, IED2 operates with settings that enable the tripping of CB2 in the case of bus fault F8 at Substation Bus-2 and facilitates the transfer trips of CB2 after receiving the CB1 open-state signal in the case of short-circuit fault F1. However, if IED2 fails to receive a CB1 open-state signal in the case of short-circuit fault F1 after the opening of CB1 and the AC microgrid already changed to islanded mode with a trip-block signal to all IEDs, this will be the failure of the transfer trip. In this case IED2 can provide a backup operation by opening CB2 with the fault current still coming from the DGs within the AC microgrid. This can be performed by IED2 either with the islanded mode settings or using the current magnitude comparison and the direct transfer trip failure logic, as explained later in Section 4. The IED2 may take quite some time to change its active group settings from the grid-connected mode settings to the islanded mode settings, and this will require DGs to remain online for additional time beyond the standard LVRT curve until the IED2 settings are changed and the tripping of CB2 is executed. However, the backup operation of CB2 by the direct transfer trip failure logic implemented at IED2 could be performed within the standard LVRT curve. In the islanded mode, the IED2 settings are adapted so that the fault F1 is detected when the CB1 is open. The 1-km cable line between Substation Bus-2 and Substation Bus-3 is protected by CB6 and CB7 with the related protection IEDs. The protection IED6 and IED7 are also considered to be adaptive.

The adaptive IED6 primarily protects both the cable line and the 20/0.4-kV transformer from short-circuit fault F2 during both the grid-connected and islanded mode of operations. In the islanded mode, after sensing the fault current at its location, the adaptive IED6 trips CB6 and transfer trips circuit breaker CB7. Additionally, IED6 and IED7 can compare the post-fault magnitude of currents at their locations with a 1.2-p.u. threshold and determine the location and direction of the fault between IED6 and IED7, as explained in the coming section. The adaptive IED7 can also provide backup protection in case of transfer trip failure (like the adaptive IED2 does, as explained earlier) in the case of the fault F2 in the grid-connected mode, in addition to the normal protection against bus fault F4 at Substation Bus-3 in both the grid-connected and islanded modes by direct tripping CB7 and transfer tripping CB9. The IED7 can only provide an "adaptive trip" to CB7 for the transfer trip failure from IED6 in the case of short-circuit fault F2 in the grid-connected mode if sufficient fault current contribution from PV is available beyond the standard LVRT characteristic curve. This is because the IED7 may need quite some time to change its active group settings from the grid-connected mode settings to the islanded mode settings, and PV must remain online until IED7 settings are changed and CB7 tripping is executed. For this purpose, a new LVRT curve proposed later in this paper can be used. The 0.2-km cable connecting WTG with Substation Bus-2 is protected by CB3 with an adaptive IED3. Both MV and LV loads are also provided with adaptive IEDs (IED5 and IED8), which trip CB5 and CB8 adaptively in the case of load-side short-circuit faults F3 and F9 in both grid-connected and islanded modes of the AC microgrid.

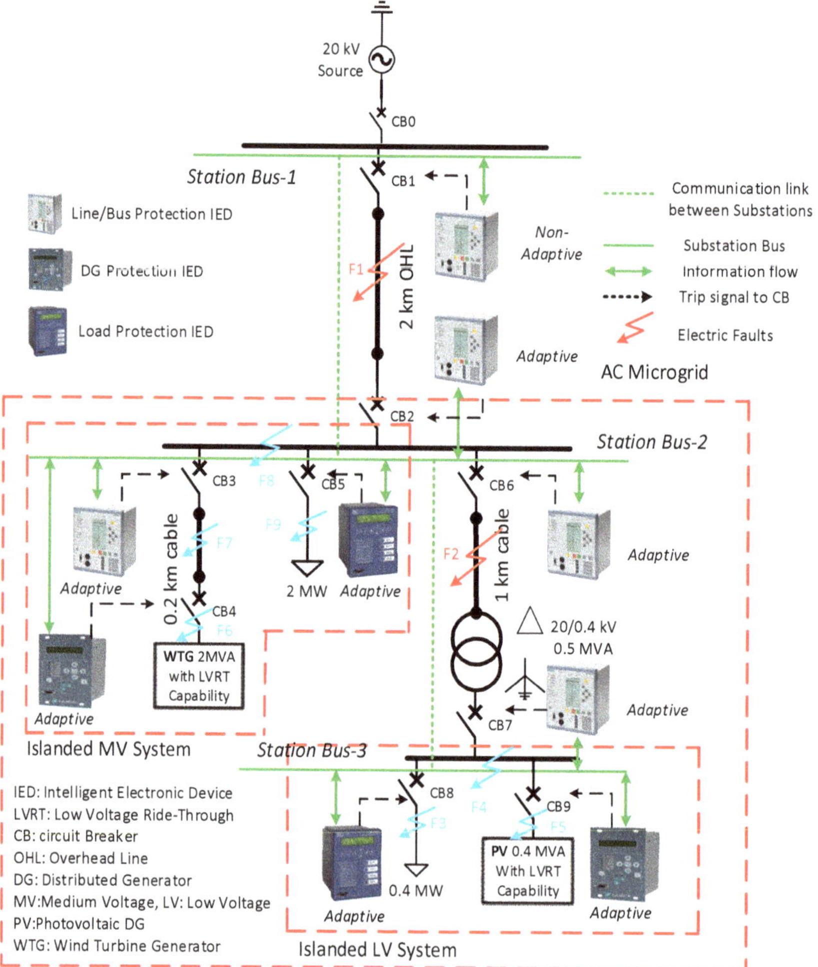

Figure 2. The radial MV/LV AC microgrid model for adaptive protection study.

The adaptive IEDs with two preplanned setting groups for AC microgrid lines are provided with only under-voltage (UV) local backup protection (Figure 3a) and adaptive IEDs with two preplanned setting groups for loads with both under and over-voltage (UV/OV) backup protection (Figure 3b). The DG protection IEDs (IED4 and IED9) are also considered to be adaptive in order to differentiate between the grid-connected and islanded mode operations. Moreover, DG protection IEDs should not trip instantaneously in the case of all external faults and allow DGs to provide fault current contributions according to predefined standard LVRT characteristics. A multifunctional adaptive IED for the protection of converter-based DGs is shown in Figure 4, which consists of adaptive OC and anti-islanding protection functions. In practice, DGs may be provided with unit protection and IEDs with several fault protection and anti-islanding protection functions. In this study, the anti-islanding protection functions (passive/active methods) of DG protection IEDs are assumed normally "disabled" if the communication link is continuous and enabled quickly when the communication link is lost. Thus, communication-based loss-of-mains detection with no nondetection zone can be used as a primary means of anti-islanding protection and passive/active methods as backup in the case of communication link failure. All the sensitive protections within the islanded AC microgrid need to be disabled/interlocked during the starting of DGs, motor loads and during the transient period when changing from the grid-connected to islanded mode and vice versa.

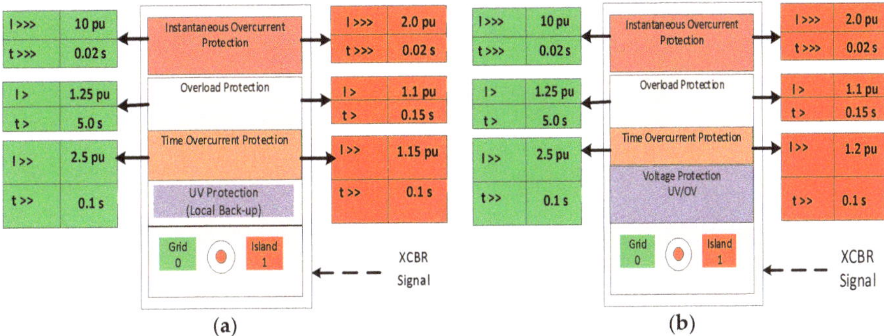

Figure 3. Adaptive definite time overcurrent (DTOC) relays with two preplanned setting groups: (**a**) for lines with local undervoltage (UV) backup protection and (**b**) for loads with local voltage protection (UV/over-voltage (OV)).

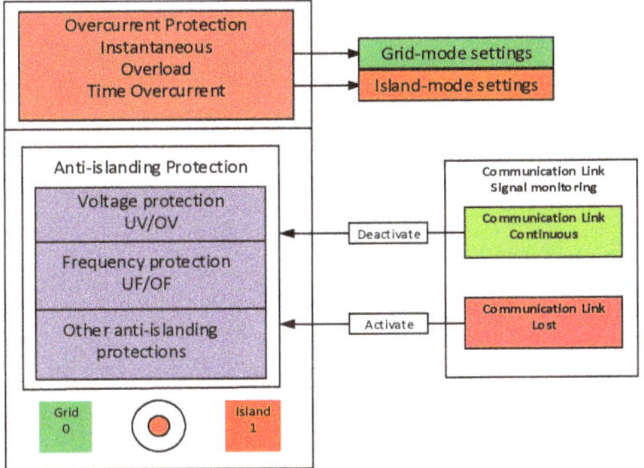

Figure 4. A multifunctional adaptive intelligent electronic device (IED) for the protection of converter-based distributed generators (DGs).

4. Adaptive Protection Methods and Results

Although several faults may happen in the presented AC microgrid, only adaptive protection methods and results of three-phase ungrounded short circuit faults with 0.01-Ohm fault resistance at locations F1 and F2 are presented. Moreover, for the sake of simplicity, it is assumed that three-phase fault F2 occurs only in the islanded mode. Nevertheless, the adaptive protection method for fault F2 also considers the protection option in the case of F2 in the grid-connected mode, as explained in the following text. Figure 5 shows the flowchart of communication-based nonadaptive IED1 for protection during fault F1. IED1 provides primary protection for fault F1 and the backup protection with definite time delay for all other downstream faults using OC relay and UV protection works as backup of the OC relay. The IED1 normally uses the redundant communication link to get information about downstream faults and use this information for trip decisions. If the fault is downstream, it waits for the CB2 to trip first. On receiving a CB2 failure signal, it trips CB1 and sends a CB1 status "open" GOOSE message (XCBR signal) to all IEDs to change their settings to the islanded mode. Even if CB1 fails, it can transfer the trip incoming breaker CB0 of substation-1 to initiate the islanding. If no communication link is available, IED1 will simply trip CB1 using definite time delays depending on the magnitude of the current. Figure 6 shows the steps for the clearance of fault F1 using GOOSE messages with different transmission delays. In both cases, at step 7, IED2 can be used in an adaptive manner for tripping CB2 to clear F1 completely, if not directly tripped by the CB1 status transfer trip, as mentioned in step 7 of Figure 6. If CB2 is not tripped with the CB1 status transfer trip or the adaptive trip by IED2, then fault F1 will not clear due to fault energization by DGs in the AC microgrid, and DGs will trip after LVRT time is elapsed.

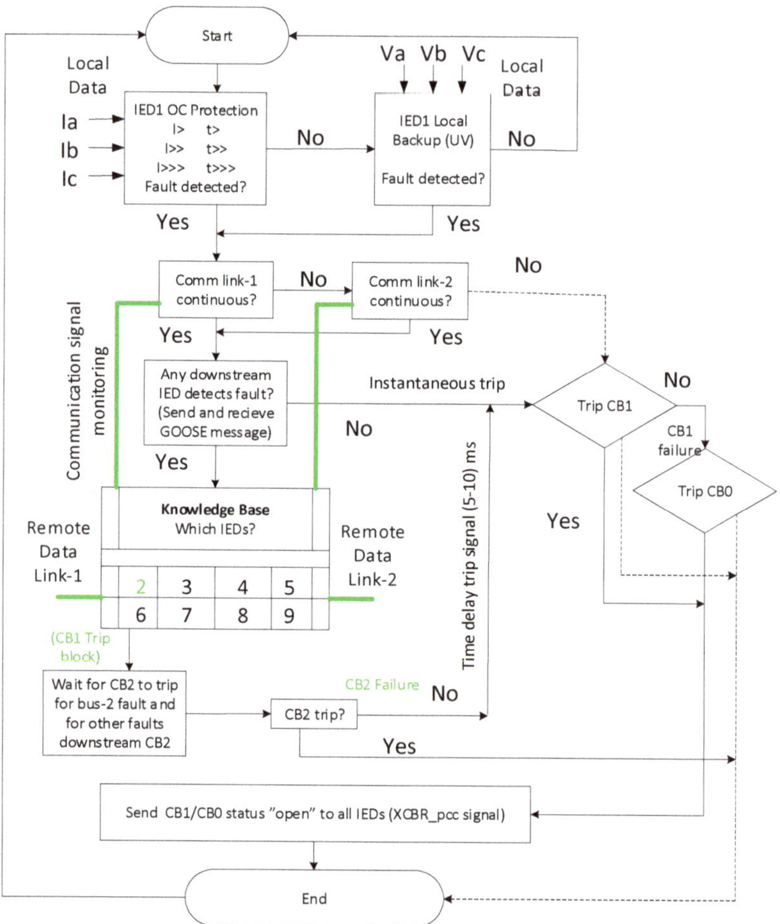

Figure 5. Flowchart for communication-based nonadaptive IED1 providing primary protection for fault F1 and remote backup for all downstream faults in the grid-connected mode. UV: Undervoltage protection.

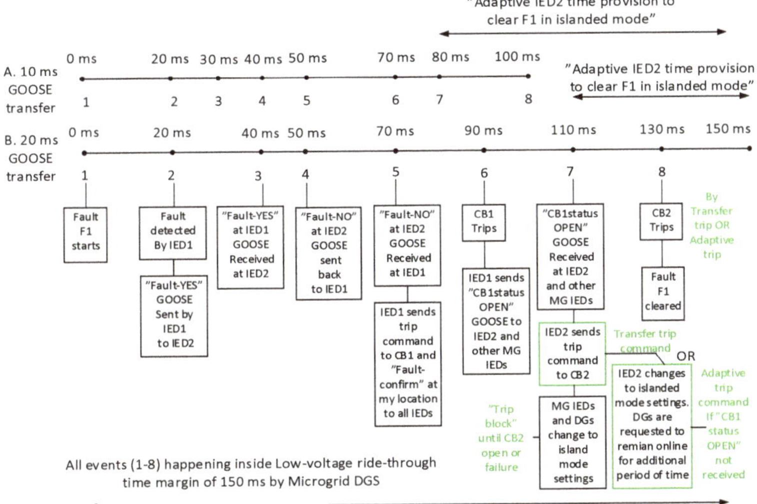

Figure 6. Fault F1 clearance time with 10-ms and 20-ms GOOSE message transfer delays (CB2 can trip by transfer trip GOOSE from IED1 or by adaptive IED2 using islanded mode settings).

Figure 7 shows the flowchart for the clearance of fault F2 in both grid-connected and islanded modes by adaptive IED6 by tripping CB6 and sending trip signal XCBR "open" to IED7 for tripping CB7. With CB6 and CB7 open, two separate islands were created within the islanded AC microgrid, one supplied by only PV (LV microgrid) and other supplied by only WTG (MV microgrid). If fault F2 occurs in the grid-connected mode, then only the LV microgrid will be isolated, and the MV microgrid will operate in the grid-connected mode. IED7 will also need the current flow direction in the case of fault F2, since this fault will be energized by both PV and WTG in islanded mode, which will avoid nuisance tripping by IED6 in the case of bus-3 fault or fault F3 at the LV load. IED7 can easily know if the fault is upstream or downstream of its location after receiving "YES fault GOOSE" from IED6 by simply calculating the RMS magnitude of the current at its location. If the magnitude of current at IED7 is ≤1.2 p.u. of the normal set current, the fault is considered to be upstream of IED7, since the fault contribution at IED7 will come from downstream PV only. In this case, IED7 will send "NO fault GOOSE" to IED6. If the magnitude of the current at IED7 is >1.2 p.u. of the normal set current, the fault is considered to be downstream of IED7. In this case, IED7 will send "YES fault GOOSE" to IED6, and IED6 will wait until the next GOOSE from IED7. The red and green colors in Figure 7 differentiate between the grid-connected and islanded mode features. On the failure of CB6, IED6 will trip CB2, CB3 and CB7 to clear F2 in the grid connected mode, whereas CB7 and CB3 will be tripped in the islanded mode to clear fault F2 completely. Hence, CB6 failure during fault F2 in both grid-connected and islanded modes will cause complete power interruptions to MV microgrid loads. Figure 8 shows the steps for the clearance of fault F2 using GOOSE messages with different transmission delays. It should be noted that, in steps 5 and 6 and 7 and 8 of Figures 6 and 8, the time delay for circuit breaker operation is considered 20 ms, which is one cycle of 50-Hz supply. This means high-speed AC circuit breakers operating in one cycle [43] will be required for the implementation of the proposed adaptive OC protection scheme.

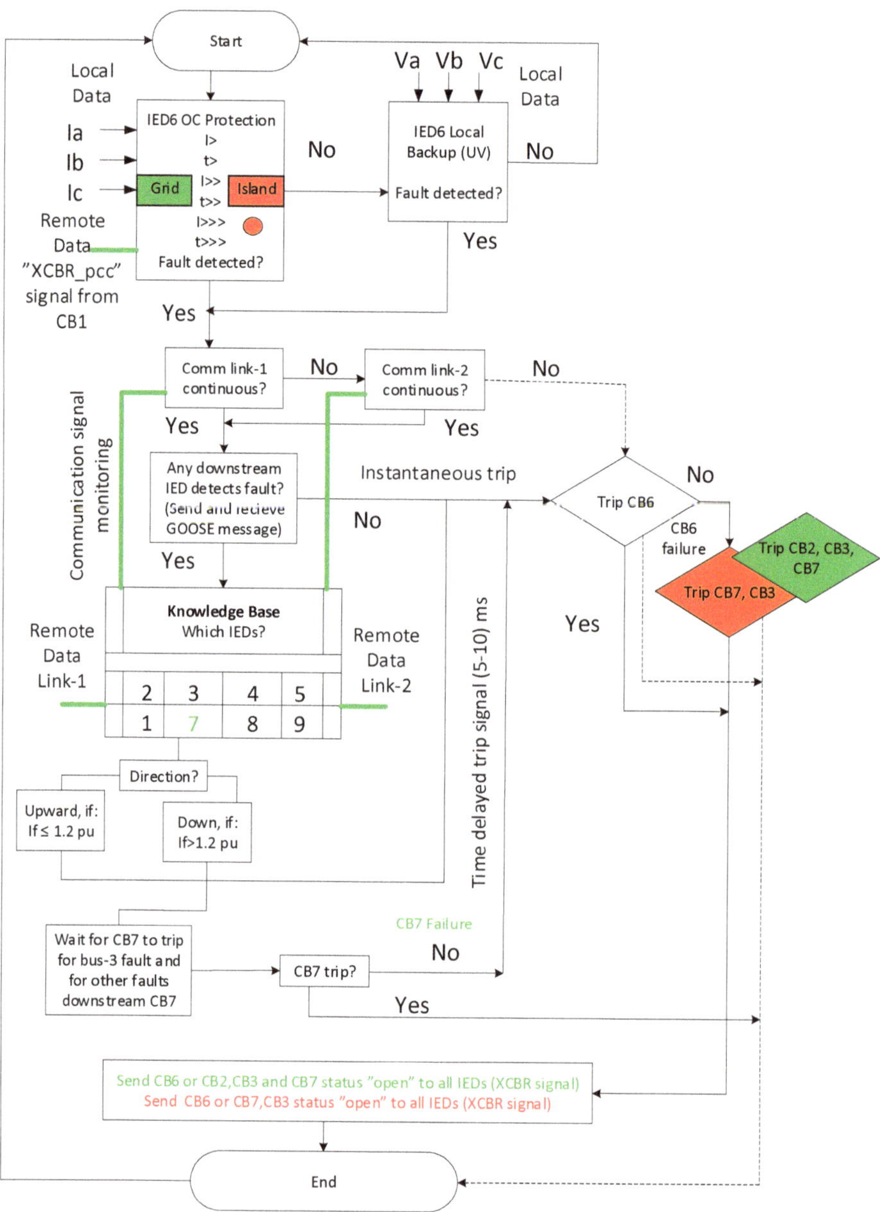

Figure 7. Flowchart for communication-based adaptive IED6 providing primary protection for fault F2 and remote backup for all downstream faults in both grid-connected and islanded modes.

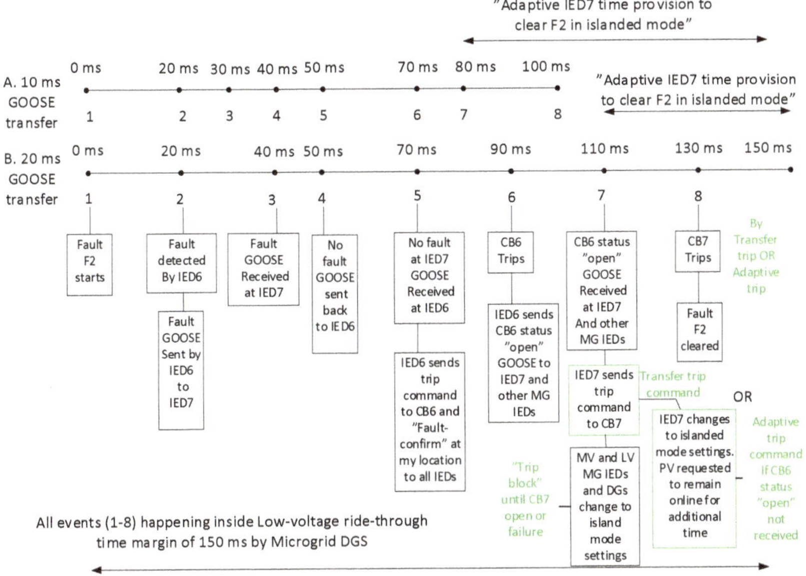

Figure 8. Fault F2 clearance time with 10-ms and 20-ms GOOSE message transfer delays (CB7 can trip by transfer trip GOOSE from IED6 or by adaptive IED7 using islanded mode settings).

Table 2 shows the normal flow of currents measured at the IED1, IED2, IED6 and IED7 locations during four different DG scenarios in the grid-connected mode. The maximum currents used for the adaptive DTOC settings are also indicated in Table 2. The fault current magnitudes at the concerned IEDs during the short-circuit fault F1 in the grid-connected mode and the short-circuit fault F2 in the islanded mode are shown in Table 3. Table 4 shows the DTOC settings and time grading of the IEDs 1, 2, 6 and 7 in the grid-connected and islanded modes. In this study, only the high-stage setting group (I>>) of Table 4 was used for the detection of the three-phase short circuit faults, and the isolation of the fault (tripping) is totally dependent on the GOOSE message transfer, according to Figures 6 and 8. However, in the case of complete communication failure, the time grading of Table 4 ensures selective operation. Table 5 explains the adaptivity and current magnitude comparison (CMC) requirements at IEDs of Figure 2 for three-phase faults (F1–F9) at different locations in different modes of operation. In the grid-connected mode, the higher setting group like that given in column two of Table 4 will be applied for all IEDs irrespective of the connection and disconnection status of DGs. This higher setting group is denoted by SG $_{GM}$ in Table 5 to represent the active settings in the grid-connected mode. For all IEDs except IED1, IED5 and IED8, a separate local current magnitude comparison function/logic is proposed for the operation in both the grid-connected and islanded modes when DGs are actively participating in load sharing with the connection status "YES". The CMC function in the IEDs will logically work in parallel with the communication-based DTOC protection and would act quickly with the "transfer trip communication failure" signal to trip the local downstream IED during the upstream fault if it does not receive the transfer trip signal (CB status "OPEN") from the upstream IED after a predefined time period (time period between step 6 and 7 in Figure 6). The CMC function will continuously receive the analog value of the current from the local MU and compare it with the current threshold of 1.2 p.u. of the max current; if the current is less than the threshold and the upstream IED also sends a fault detection GOOSE "YES", the fault is assumed to be the upstream fault. Then, if no transfer trip GOOSE is received from the upstream IED within 110 ms of the fault (event 7 in Figure 6), the IED will trip the local circuit breaker to clear the fault completely. In this way, the CMC function implemented in the communication-based adaptive protection (Figure 7) could not only detect the direction of the fault, but it could also act as a backup for direct transfer trip (DTT) communication failure. The proposed CMC function can only work for the feeder with a strong DG source on one side and a comparatively weaker DG source with a predefined fault current contribution on the other side. For example, during the fault F1 in the grid-connected mode when the connection status of both the WTG and PV is "YES", the IED2-IED4, IED6-IED7 and IED9 all will need the CMC function to determine if the fault is at the downstream or the upstream location. However, only IED2 will activate the CMC trip function after the transfer trip GOOSE message from IED1 is found undetected (not received). In the same way, during the fault F2 in the islanded mode of operation when the connection status of both the WTG and PV is "YES", the IED6-IED7 and IED9 need the CMC function to know if the fault is upstream or downstream of these relays. IED6 using the CMC function will detect the fault F2 to be downstream, while IED7 and IED9 will detect the fault to be upstream of their locations. Then, when the transfer trip GOOSE from IED6 is not received by the IED7, it will initiate the backup using CMC logic to trip CB7 in order to clear the fault F2 completely, while IED9 will continue following the LVRT curve. The topmost AND logic in Figure 9a,b presents the backup for DTT failure from IED6 during the fault F2 using the CMC function/logic. It should be noted that, in the islanded mode, when the connection status of both the WTG and PV is "YES"(the scenario of Table 5, column 3), then the WTG of 2 MW acts as a comparatively stronger source than the PV of 0.4 MW. Therefore, the grid-connected mode settings "SG $_{GM}$" for IED6-IED9 will remain effective and unchanged in this islanded mode scenario, and only IED3-IED5 need to change to the islanded mode settings (SG $_{IM}$). This way, the minimum number of IEDs will need to adapt to the islanded mode settings, and simple logics for the primary and backup protections, for example, at IED7 (Figure 9a), could be used. This will also prevent the OC function of IED7 and IED9 to start pickup for the "out of zone" fault F2 in the islanded mode scenario of Table 5, column 3.

Table 2. Normal RMS (Root mean square) currents at considered intelligent electronic device (IED) locations with four different distributed generator (DG) scenarios in grid-connected mode. WTG: wind turbine generators and PV: photovoltaic.

IED No.	Current (A) Only Grid Scenario-1	Current (A) WTG + Grid Scenario-2	Current (A) PV + Grid Scenario-3	Current (A) WTG + PV + Grid Scenario-4
IED1	69.1 [1]	12.476	58.39	1.098
IED2	69 [1]	12.43	58.34	1.05
IED6	11.571	11.572 [1]	2.873	2.876
IED7	573.8	575 [1]	182.81	182.84

[1] The maximum currents used for adaptive overcurrent (OC) settings.

Table 3. Fault currents (RMS) at considered IED locations with DG scenario-4 (F1 in grid-connected mode and F2 in islanded mode).

IED No.	Current (A) during Fault F1 in Grid-Connected Mode	Current (A) during Fault F2 in Islanded Mode
IED1	28,000	-
IED2	67 [2]–82.8 [3]	-
IED6	-	70
IED7	-	700

[2] Before CB1 tripping (grid-connected mode). [3] After CB1 tripping (used for IED2 adaptive tripping).

Table 4. Settings of overcurrent IEDs for grid-connected and islanded modes as per DG scenarios 1–4 in Table 2 considering the max normal magnitude of current through IEDs in these scenarios.

IED No.	DTOC Relay Settings for Grid-Connected Mode (SG $_{GM}$)	DTOC Relay Settings for Islanded Mode (SG $_{IM}$)
IED1	$I\!\!>> 2.5 \times 69.1 = 172.75$ [4] A $t\!\!>> 0.8$ [6] s	-
IED2	$I\!\!>> 2.5 \times 69 = 172.5$ [4] A $t\!\!>> 0.6$ [6] s	$I\!\!>> 1.15 \times 69 = 79.35$ [5] A $t\!\!>> 0.6$ [6] s
IED6	$I\!\!>> 2.5 \times 11.572 = 28.93$ [4] A $t\!\!>> 0.4$ [6] s	$I\!\!>> 1.15 \times 11.572 = 13.31$ [5] A $t\!\!>> 0.4$ [6] s
IED7	$I\!\!>> 2.5 \times 575 = 1437.5$ [4] A $t\!\!>> 0.2$ [6] s	$I\!\!>> 1.15 \times 575 = 661$ [5] A $t\!\!>> 0.2$ [6] s

[4] Definite time overcurrent (DTOC) setting group for the grid-connected mode (SG $_{GM}$). [5] DTOC setting group for the islanded mode (SG $_{IM}$). [6] Conventional time-coordination only used in the case of communication failure.

Table 5. The adaptivity of SG $_{GM}$ or SG $_{IM}$ settings and CMC [8] requirements at various IEDs (Figure 2) during different faults in grid-connected and islanded mode scenarios with predefined fixed fault current contributions from DGs.

IED No.	Grid-Connected Mode (WTG + PV)-Yes	Islanded Mode (WTG + PV)-Yes	WTG-No, PV-Yes PV-Export Mode	PV Island
IED1	SG $_{GM}$ (F1–F9) [8]	-	-	-
IED2	SG $_{GM}$ + CMC [7] (F1)	-	-	-
IED3	SG $_{GM}$ + CMC (F1, F8 and F9)	SG $_{IM}$ + CMC (F8 and F9)	-	-
IED4	SG $_{GM}$ + CMC (F1, F7–F9)	SG $_{IM}$ + CMC (F7–F9)	-	-
IED5	SG $_{GM}$ (F1–F9)	SG $_{IM}$	SG $_{IM}$	-
IED6	SG $_{GM}$ + CMC (F1, F8 and F9)	SG $_{GM}$ + CMC (F2, F8 and F9)	SG $_{IM}$	-
IED7	SG $_{GM}$ + CMC (F1–F2, F7–F9)	SG $_{GM}$ + CMC (F2, F7–F9)	SG $_{IM}$	-
IED8	SG $_{GM}$ (F1–F9)	SG $_{GM}$	SG $_{IM}$	SG $_{IM}$
IED9	SG $_{GM}$ + CMC (F1–F4, F8 and F9)	SG $_{GM}$ + CMC (F2 and F3, F8 and F9)	SG $_{IM}$	SG $_{IM}$

[7] Current magnitude comparison: $I_{IED} \leq 1.2$ p.u. of max possible normal current at a location, then fault is upstream; $I_{IED} > 1.2$ p.u. of max possible normal current at a location, then fault is downstream. [8] Fault locations (F1 to F9) for which the CMC feature is required at the corresponding IED. Note: Table 5 considers the minimum required settings of the main (primary) protection for the detection of the faults, and the final successful islanded scenarios are considered.

If, during the islanded mode (Table 5, column 3 mode), the settings of IED6, IED7 and IED9 are changed to the islanded mode settings (SG $_{IM}$), then OC function of all the three IEDs will start picking up not only during the downward fault F5 but, also, during the upward faults F4 and F2. Hence, the protection logic for the detection of the "in-zone" fault direction and location will become more complex with the islanded mode settings (Figure 9b) compared with the protection logic with the grid-connected mode settings (Figure 9a). The bottom AND logic of Figure 9a represents the protection logic of IED7 with the grid-connected mode settings (SG $_{GM}$) for "in-zone" fault F4, where only "NO fault" detection signals at downward IED8 and IED9 are sufficient to activate the primary and time-delayed backup protection at IED7, in addition to the local IED7 and upward IED6 "fault YES" signals and the CMC function outputs for "downward" fault. The bottom AND logic (black) of Figure 9b represents the protection logic of IED7 with the islanded mode settings (SG $_{IM}$) for the "in-zone" fault F4, where not only "NO fault" detection at downward IED8 and IED9 are required, but also, the CMC function output at IED9 for "upward" and "downward" faults will also be required to activate the primary and time-delayed backup protection at IED7, respectively, in addition to the local IED7 and upward IED6 "fault YES" signals and the CMC function outputs for the "downward" fault. The protection logics presented in Figure 9 will be necessary in order to quickly detect the fault and isolate the faulty section within 150 ms, keeping the stability of the remaining system intact if the communication system performance is according to the predefined boundaries. In the case of communication failure, the normal communicationless time coordination described in Table 4 will be applied.

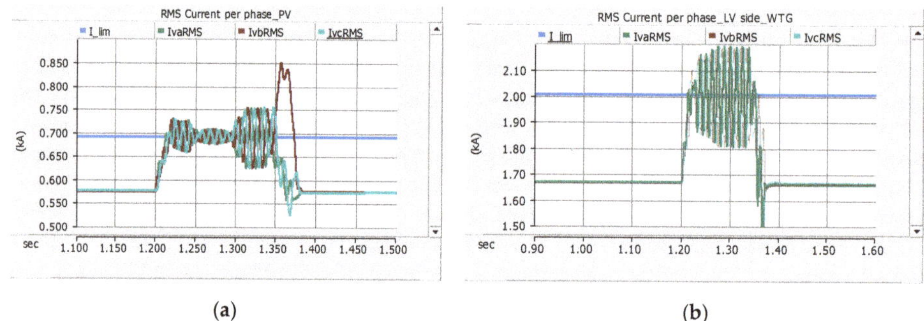

Figure 12. RMS magnitude of current per phase of DGs before, during and after fault F1 (CB2 transfer trip): (**a**) LV side of the PV system and (**b**) LV side of WTG. Note: I_lim is the preset limit of the fault current contribution of DGs.

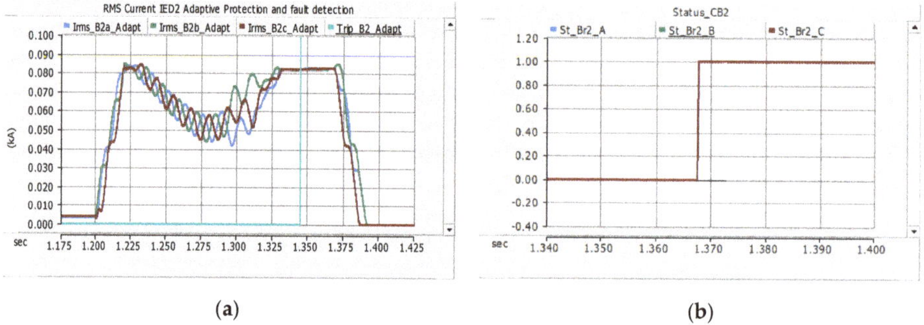

Figure 13. Adaptive IED2 trip after the unsuccessful direct transfer trip from IED1: (**a**) RMS magnitude of the current at adaptive IED2 and relay operating time and (**b**) CB2 status and operating time during an IED2 adaptive trip.

Figure 14 shows the RMS magnitude of the current of DGs before, during and after the fault F1 with an adaptive IED2 trip. The results in Figure 13 show that it takes 20 ms more than the required 150 ms time for clearance of the fault with adaptive IED2 settings. This is because the DG output current takes an extra 10 ms to stabilize and reach the threshold setting of adaptive IED2. Moreover, both IED2 and CB2 take 5 ms more than the set time of 20 ms for fault detection and tripping. Figure 13 also shows that it will take 15 ms extra for DGs to restore to normal operations (normal currents) in the islanded AC microgrid after the removal of fault F2 completely at 1.37 s. These types of extra delays highlight the demand of faster communication with less than 20 ms one-way transfer delay or extension of the initial time after the fault inception in the LVRT characteristic curve for effective adaptive protection in order to maintain the supply in islanded mode. Otherwise, it will cause the complete loss of DGs within the AC microgrid during fault F1 due to the tripping by anti-islanding (e.g., UV) protection at 150 ms after the fault. This, of course, will decrease the reliability of supply for AC microgrid loads due to extra time for the black start of DGs in the islanded mode, then removal of fault F1 by adaptive IED2 and, finally, the restoration of the normal load. In the presented cases by using a 20-ms delay of one-way GOOSE transfer for the complete removal of fault F1, the transfer trip method is faster than using adaptive IED2 tripping in the islanded mode: a direct transfer trip takes about 150 ms after the fault, and the second backup procedure (adaptive tripping) takes about 190 ms after the fault. However, if the delay of 10 ms for one-way GOOSE transfer is used (Method-A in Figure 6), then even the backup adaptive tripping after the failure of the direct transfer trip could be accomplished within 150 ms after the fault, and the present LVRT curve of the DGs will remain valid.

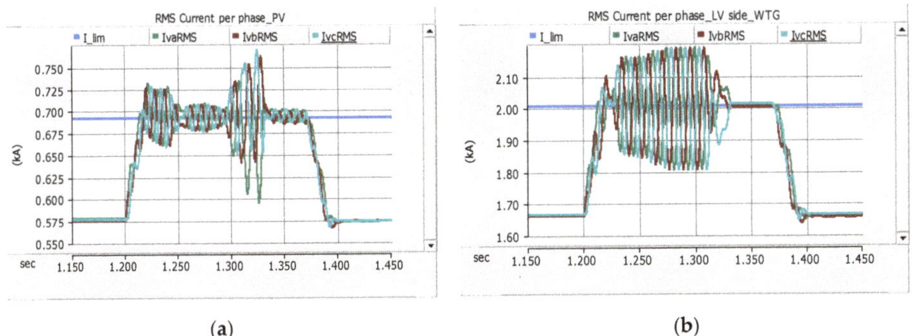

(**a**)   (**b**)

Figure 14. RMS magnitude of the current per phase of DGs before, during and after fault F1 (adaptive IED2 trip): (**a**) LV side of the PV system and (**b**) LV side of the WTG. Note: At 1.29–1.3 s, CB1 trips (Figure 10a), and at 1.37 s, CB2 adaptively trips (Figure 12b).

Control of DGs

Both DGs (WTG and PV) are operated in fixed P-Q (active power-reactive power) control near the unity power factor operation; their active and reactive power supply before, during and after the fault F1 are shown in Figure 15. In grid-connected mode, the current-controlled voltage source converters (VSCs) of both the WTG and the PV models operate in the grid-imposed frequency mode (the grid-following mode) using the PLL (phase locked loop). However, after receiving the CB1 "open" signal, the current-controlled VSCs of both DGs use the reference voltage angle from the free-running VCO (voltage controlled oscillator) and operate in the controlled-frequency mode (the grid-forming mode), since the PLL loses its synchronism after the loss of grid voltage. The variation of the LV side currents of DGs observed during the fault F1 in the grid-connected mode before the tripping of CB1 and before the activation of VCO (Figures 12 and 14) are due to PLL errors in the simulation model during the fault, but this does not cause errors in the OC function operations. Some additional resistances (0.39 ohm per phase) are connected in the series with the output terminal inductors of the PV system to maintain a terminal voltage constant in islanded mode. These types of grid-following and grid-forming VSC controls are explained in more detail in [44]. Previously, the conventional f/P (frequency/active power) and V/Q (voltage/reactive power) droop is applied to at least one grid-side converter of a WTG from the group of WTGs connected to the same bus to act as the voltage and frequency control sources in the islanded mode. The converters of the remaining WTGs follow the controlled system frequency. In this way, the response of the islanded system due to the sudden large changes of the load is kept smoother compared with the controls where the voltage and frequency droops are applied to converters of all WTGs [45,46]. The results in Figures 16 and 17 indicate how the AC microgrid is smoothly transitioned to the islanded mode with the applied controls of DGs after the fault F1 is cleared. Figure 16 indicates the variation of frequency during the clearance of fault F1; therefore, the frequency immunity of DGs is also required in addition to the standard LVRT characteristics.

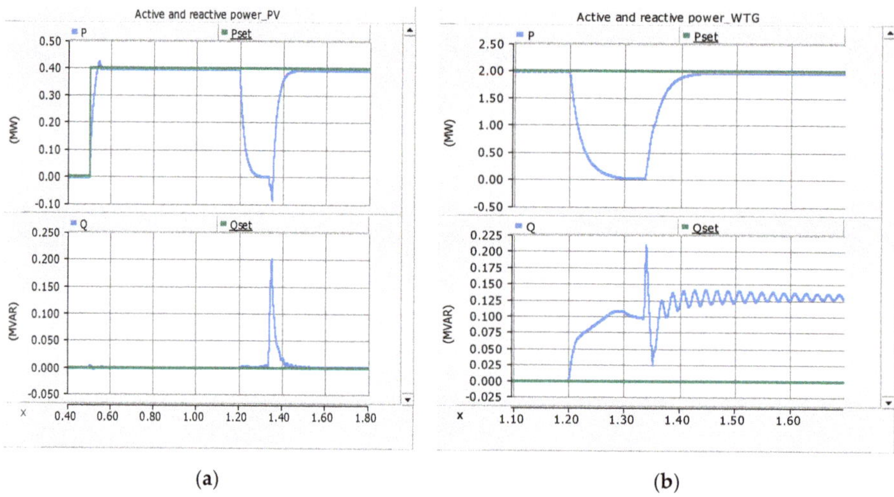

(**a**)   (**b**)

Figure 15. T Active power (P) and reactive power (Q) supply from DGs before, during and after fault F1: (**a**) P and Q from the PV system and (**b**) P and Q from the WTG. Note: WTG supplies the entire Q demand in islanded mode.

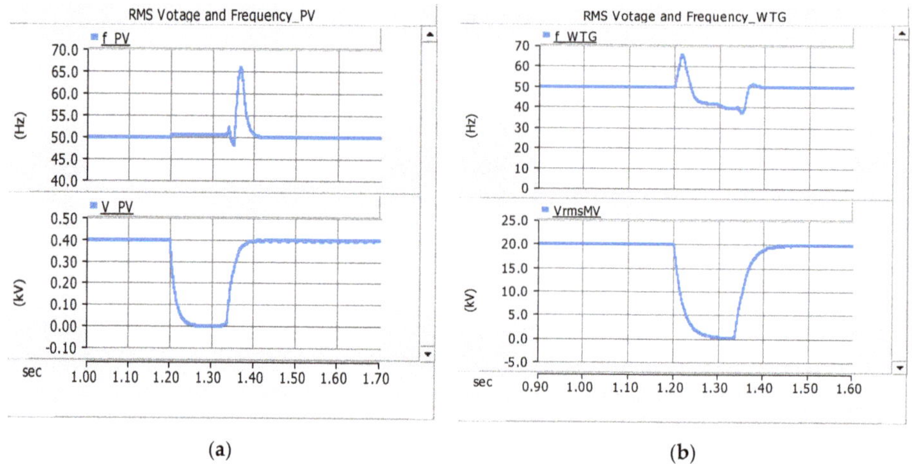

(**a**)   (**b**)

Figure 16. Three-phase RMS voltage and frequency of DGs before, during and after fault F1 (**a**) low-voltage (LV) side of the PV system and (**b**) medium voltage (MV) side of the WTG.

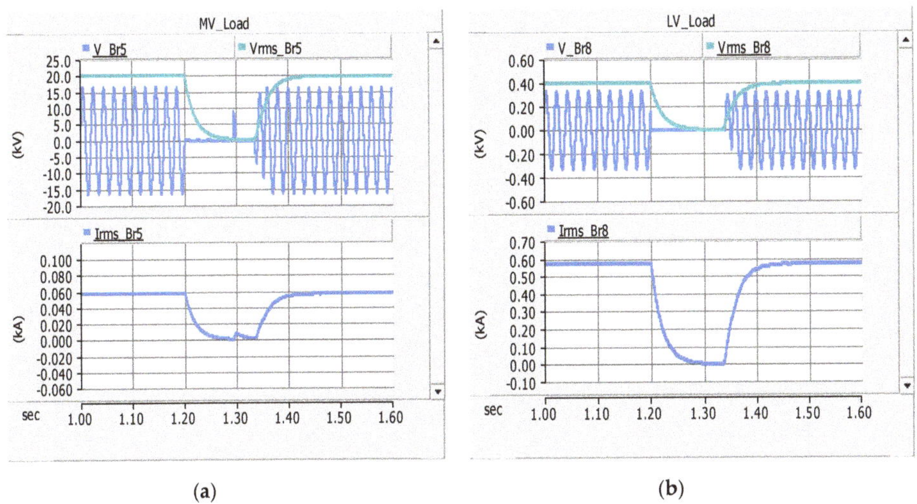

(a)   (b)

Figure 17. Three-phase RMS voltage and current and single-phase instantaneous voltage at loads before, during and after fault F1: (**a**) MV load and (**b**) LV load.

4.2. Results for Fault F2 in Islanded Mode and the Creation of Two Islands within the Islanded AC Microgrid

For three-phase fault F2 in the islanded mode (CB1 and CB2 open in Figure 2), the adaptive OC relay logics are implemented in PSCAD according to Figure 8 and settings according to Figure 3 and Table 4. The fault starts at 1.2 s and ends at 5 s; this fault duration is small, but it is assumed to be a permanent fault. The fault current magnitude at IED6 and IED7 before, during and after fault F2 in the islanded mode is shown in Figure 18. It shows that the fault current magnitude is enough at the IED6 location, which is supplied by the WTG that is set to supply 1.2 p.u. of its rated current during the fault. The fault current at IED6, which is supplied by the WTG, is considerably higher than the maximum current at IED6 during any DG scenario, and therefore, the fault can be detected easily even with the grid-connected mode higher OC settings (SG $_{GM}$) of IED6 (Table 4). Hence, IED6 can be a nonadaptive IED for this fault case. The fault current magnitude is limited at the IED7 location during F2, which is supplied only by the PV system; therefore, the fault F2 can only be detected by IED7 with islanded mode lower settings (SG $_{IM}$). The IED7 should be necessarily adaptive in order to work even when the transfer trip from IED6 fails. Since the AC microgrid in this case is islanded, the settings of IED7 are already changed to islanded mode settings; hence, both IED6 and IED7 can detect the fault and trip simultaneously to remove the fault F2 after checking the magnitude of the current at downstream IED8. Alternatively, the trip block signal can be issued to IED7 from IED6, and IED7 can later be transfer-tripped after the opening of CB6. The results shown in this section are based on IED6 with one setting group (SG $_{GM}$) that can detect the fault F2 in both the grid-connected and islanded modes. Although IED7 is adaptive, it has been transfer-tripped by IED6 in these results (Figure 19), according to method-B (20-ms GOOSE transfer) of Figure 8. Figure 20 shows the RMS currents of DGs before, during and after the fault F2 in islanded mode with fault clearance using CB2 direct transfer trip. Figure 21 shows the active and reactive power supply from DGs before, during and after the fault F2. Additionally, it is shown in Figures 22 and 23 how smoothly two islands are formed within the islanded AC microgrid after the clearance of the fault F2 at 1.34 s. The results of adaptive IED7 tripping after transfer trip failure from IED6 during the fault F2 are not included to avoid repetition, in which case the adaptive IED7 may detect the fault with the lower settings (SG $_{IM}$) just like the adaptive IED2, as explained in Section 4.1. In that case, the adaptive IED7 will wait until the time of direct transfer trip is elapsed; this is considered as a transfer trip failure from IED6. The adaptive IED7 will then decide to trip CB7 with the lower settings for the complete clearance of the fault F2 during transfer trip failure.

It should also be noted that the WTG is comparatively stronger source than the PV system. Therefore, for faults F3, F4 or F5 in islanded AC microgrid (CB1 and CB2 open in Figure 2), the WTG may still provide sufficient fault current, and the higher current settings (SG $_{GM}$) for IED7, IED8 and IED9 may still work for any of the faults F3–F5 downstream of the WTG. For these faults (F3–F5), the current comparison method to find the location of the faults (upstream or downstream faults) will also be valid for the islanded mode with the WTG in operation. However, after the removal of the fault F2 (CB6 and CB7 open), two further islands will be created: the islanded MV system and the islanded LV system (Figure 2). The islanded MV system will be supplied by only the WTG, and the islanded LV system will be supplied by only the PV system. In this situation, only the adaptive lower settings (SG $_{IM}$) of IED8 and IED9 will work. For any islanded scenario, IED3, IED4 and IED5 will always require lower adaptive settings (SG $_{IM}$).

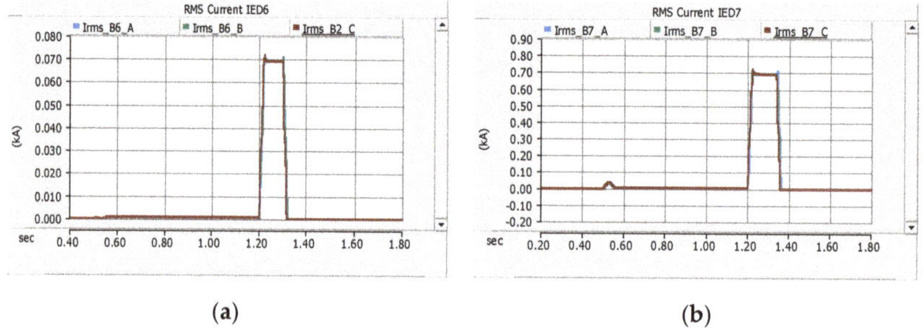

(a)

(b)

Figure 18. RMS current magnitude before, during and after fault F2 in the islanded mode at: (**a**) IED6 and (**b**) IED7 (IED7 direct transfer trip).

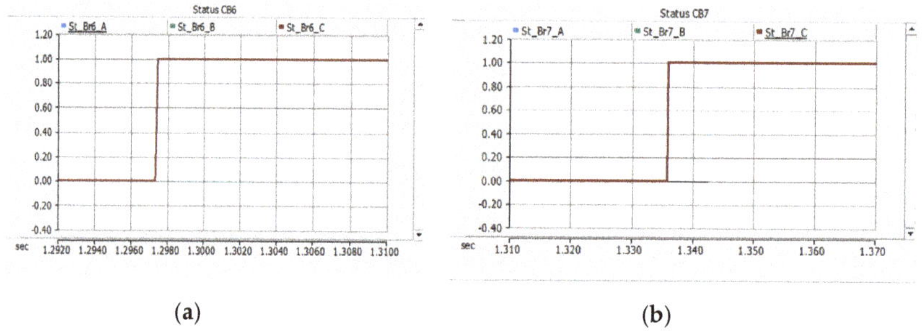

(a)

(b)

Figure 19. The operating times and status of breakers before, during and after the clearance of fault F2 in the islanded mode: (**a**) CB6 (main protection trip) and (**b**) CB7 (direct transfer trip).

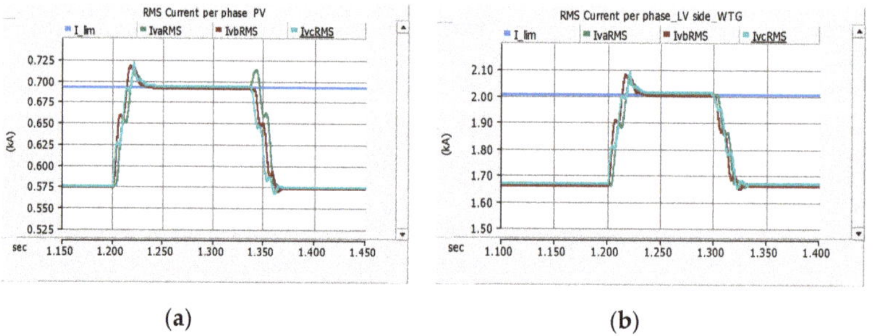

(a)

(b)

Figure 20. RMS current magnitude per phase of DGs before, during and after fault F2 in the islanded mode (CB7 transfer trip): (**a**) LV side of the PV system and (**b**) LV side of the WTG.

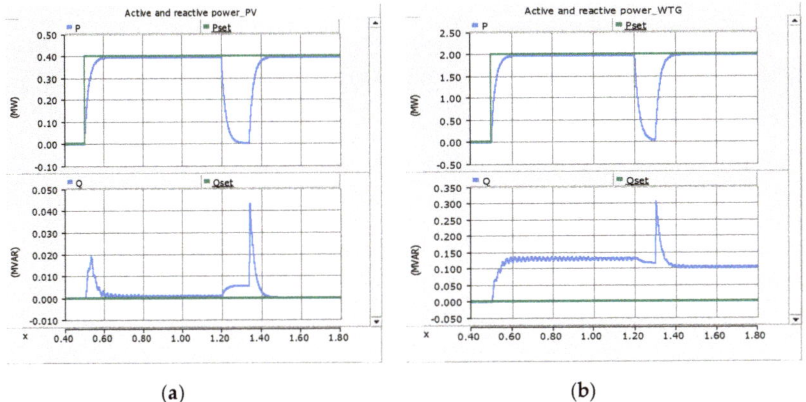

Figure 21. Active (P) and reactive (Q) power supply from DGs before, during and after fault F2 in the islanded mode: (**a**) P and Q from the PV system and (**b**) P and Q from the WTG.

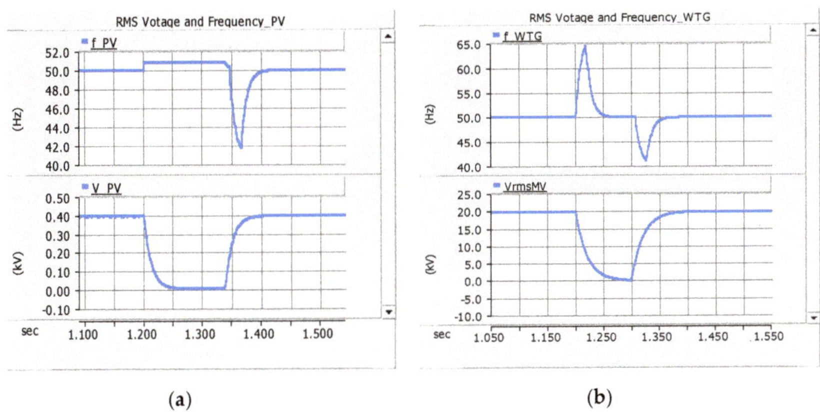

Figure 22. Three-phase RMS voltage and the frequency of DGs before, during and after fault F2 in islanded mode at: (**a**) the LV side of the PV system and (**b**) MV side of the WTG.

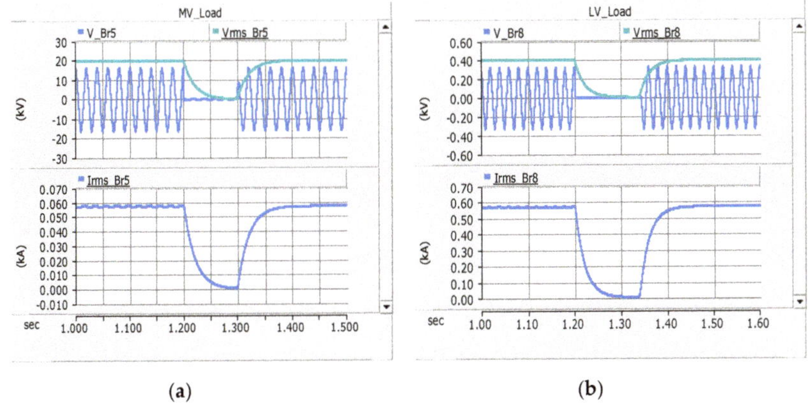

Figure 23. Three-phase RMS voltage, current and single-phase instantaneous voltage at loads before, during and after the fault F2 in the islanded mode at: (**a**) the MV load and (**b**) LV load.

The LVRT Capability of DGs

The DGs are required to remain connected to the network during the voltage dips or faults, and depending on the requirements, the DGs may be requested to feed-in short-circuit fault currents up to the agreed value (or according to the limits of DGs) in order to support the detection of faults. This is called the low-voltage ride through (LVRT), under-voltage ride through (UVRT) or fault ride through (FRT) requirement. In the grid-connected mode, the LVRT capability of DGs is usually required to maintain the stability of the system, since the disconnection of many DGs even for a fraction of a second may result in large voltage or frequency fluctuations, causing the voltage or frequency instability of the entire system. The synchronous generators are more sensitive to voltage dips, and hence, their LVRT requirements are considerably less stringent in comparison with the nonsynchronous generators, including the converter-based generators, which can remain connected for the extended durations. In the islanded mode of operation, not only the system stability is important, but also, the quick detection and isolation of the fault is equally important. Additionally, the protection coordination or selectivity between the main and backup protection has to be ensured. Figure 24 shows the comparison of different LVRT requirements of nonsynchronous generators, including the converter-based generators, according to the previous German BDEW-2008 standard [32] and the latest European Standard EN 50549-1:2019 adopted as the Finnish National Standard SFS-EN 50549-1:2019 [47]. The DGs are required to remain connected in parallel with the LV or MV networks if the voltage at the connection point is above the voltage-time curves of Figure 24 (red and black curves). Although, the LVRT requirements are expectedly limited to the most stringent curves, however, the network operators may define their own LVRT characteristics. These standards do not define the LVRT requirements for the islanded mode of operation. Therefore, a new LVRT characteristic was proposed in this paper for the islanded mode operation of the nonsynchronous DGs, including the converter-based DGs shown as a green voltage-time curve in Figure 24. According to this new proposed LVRT characteristic, the DGs will remain connected to the islanded MV/LV microgrid for at least 2 s after the voltage dip or the fault and feed-in short-circuit current of at least 1.2 p.u. of the rated current. With the proposed LVRT characteristic, not only the stability of the islanded microgrid will be maintained, but also, a good protection coordination between adaptive IEDs will be ensured. The WTG and BESS (battery energy storage systems) with full-scale converters are capable of providing this requirement. Normally, in the presence of high-speed communication, the standard grid-connected LVRT curves will be used; however, in case of communication failure, the definite-time coordination of IEDs with the proposed LVRT curve will be applied.

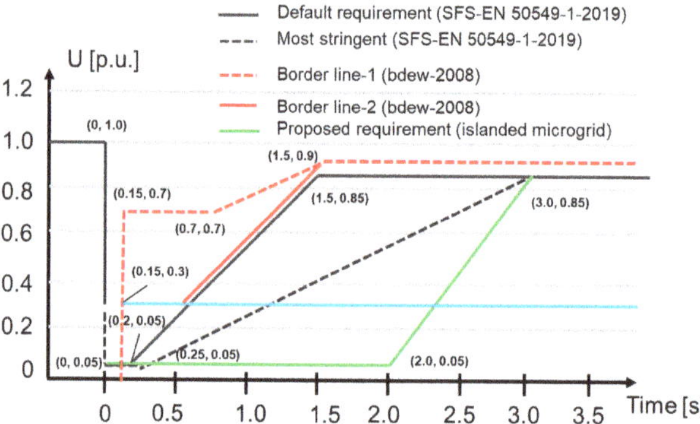

Figure 24. The low-voltage ride through (LVRT) capability of the nonsynchronous generators, including the converter-based generators: The red and blue voltage-time curves according to [32] and the black voltage-time curves according to [47]; the green voltage-time curve is the new proposed LVRT curve for the islanded mode of operation.

5. Discussion

An adaptive OC protection for the AC microgrid using the IEC 61850 communication standard and LVRT capability of DGs is presented in this paper. Previously, an adaptive OC protection was presented in [48]; the scheme updated the inverse-time OC relay curve by changing the pick-up current with respect to DG infeed, but the focus was mainly on medium impedance faults at the end of the radial distribution network detected by a single-substation relay. The similar type of adaptive OC protection was proposed in [49] for distribution networks with DGs using local data and two-setting groups for the inverse-time OC relay. However, the scheme does not use remote data by communication systems and is prone to nuisance and slow trips, resulting in DG loss. A directional adaptive inverse-time relay was presented more recently for HIL (Hardware-In-Loop) testing, and real-time simulations in [50,51]. In these papers, an FCL was used for limiting the wind turbine generator fault contribution, and its effects on relay settings were observed. This directional adaptive inverse-time relay using a multiagent system does not consider/mention the effects of communication delays on protection coordination. Moreover, it involves tedious calculations to generate various inverse time setting groups for changing the network configurations and those for all relays in the networks. Although, traditionally, inverse-time OC performs better than definite-time OC in terms of the minimum operation time close to the power source, but in the AC microgrid environment with many DG sources, this may not be completely true. The inverse-time OC relays are affected the most with the increased penetration level of DGs [4], and it is the usual practice to limit the fault contribution from DGs to overcome the adverse effects, as it was also done in [51]. An adaptive OC protection for distribution networks was proposed in [52] that calculated and applied the new settings of OC relays directly whenever any significant change in the network occurred. The algorithm presented did not use precalculated settings and was initiated either by the monitoring block in the coordination layer or energy management system during topology changes using the communication link. The scheme was verified using the real-time digital simulator (RTDS) and IEC 61850 GOOSE messaging. However, the adaptive OC scheme was implemented using a centralized approach; the type of DG unit was not specifically described, and higher coordination delays were used. Moreover, the proposed adaptive OC scheme also gave slower tripping times compared with the traditional OC relay in some cases. In this paper, the communication-based definite-time OC relays with only two predefined setting groups are suggested. These two setting groups can be changed adaptively and quickly by each IED autonomously after receiving the status of DGs (on/off), CB status change (open/close), fault current magnitude and fault detection GOOSE signal from other IEDs. Moreover, the simple method of calculating the magnitude of the current at their locations, comparing it with the predefined threshold 1.2 p.u. of the max current and sharing this information with other IEDs will be useful for quick detection of the fault location after knowing the status of each DG. The nuisance tripping can also be avoided with fault current magnitude sharing between IEDs and with the careful use of trip block/release signals. The proposed current magnitude comparison method also avoids the additional measurement of voltage for the detection of the fault current direction. However, a voltage magnitude measurement can be used as a local backup protection for OC function. The proposed scheme in this paper can also be extended to low impedance single-phase ground faults and other asymmetrical faults. The method proposed in this paper will be evaluated with a real-time digital simulator of OPAL-RT for hardware-in-the-loop (HIL) simulations using actual Ethernet-based GOOSE communication between IEC 61850-based IEDs from different vendors for its practical implementation, and this will be presented in a separate research article in the future.

6. Conclusions

The adaptive OC protection utilizing the LVRT characteristic of DGs and using circuit breaker status signals transmission by IEC 61850 communication standard were presented for the grid-connected and the islanded mode of the radial AC microgrid. Moreover, a new LVRT curve for the islanded mode of operation is proposed. The fixed delays for GOOSE message communication between IEDs are assumed according to IEC 61850 and the practical values. According to the considered assumptions, the results look promising, as per the evaluation by PSCAD simulations. The general methodology presented in this paper for three-phase faults can also be extended to other types of faults. The effectiveness of the considered adaptive OC protection for single-phase and high impedance faults for AC microgrids with different grounding schemes will be an important and interesting topic of the future study. Moreover, HIL simulations using actual IEDs and Ethernet-based IEC 61850 communication will also be performed for the practical implementation of the proposed methods.

Author Contributions: Conceptualization, A.A.M.; methodology, A.A.M.; software, A.A.M.; validation, A.A.M.; formal analysis, A.A.M.; investigation, A.A.M.; data curation, A.A.M.; writing—original draft preparation, A.A.M.; writing—review and editing, K.K.; visualization, A.A.M.; supervision, K.K.; project administration, K.K. and funding acquisition, K.K. All authors have read and agreed to the published version of the manuscript.

Funding: This work was carried out mainly in the Protect-DG research project with the financial support provided by the European Regional Development Fund (ERDF) through Business Finland with Grant No.4332/31/2014. Some parts of this work were done in the SolarX research project with the financial support provided by the Business Finland with Grant No. 6844/31/2018. The financial support provided through these research projects is highly acknowledged.

Acknowledgments: The corresponding author is very thankful to his colleague Sampo Voima for useful discussion and collaboration about the topic during the earlier stage work of the paper in the Protect-DG research project.

Conflicts of Interest: The authors declare no conflict of interest.

References

1. Microgrids at Berkeley Lab, Grid Integration Group. Energy Storage and Distributed Resources Division, Microgrids, Microgrid Definitions. Available online: https://building-microgrid.lbl.gov/microgrid-definitions (accessed on 30 June 2020).
2. Justo, J.J.; Mwasilu, F.; Lee, J.; Jung, J.W. AC-microgrids versus DC-microgrids with distributed energy resources: A review. *Renew. Sustain. Energy Rev.* **2013**, *24*, 387–405. [CrossRef]
3. Li, X.; Dyśko, A.; Burt, G. Enhanced protection for inverter dominated microgrid using transient fault information. In Proceedings of the 11th IET International Conference on Developments in Power Systems Protection (DPSP 2012), Birmingham, UK, 23–26 April 2012; pp. 1–5.
4. Memon, A.A.; Kauhaniemi, K. A critical review of AC Microgrid protection issues and available solutions. *Electr. Power Syst. Res.* **2015**, *129*, 23–31. [CrossRef]
5. Oudalova, A.; Fidigatti, A. Adaptive Network Protection in Microgrids. *Int. J. Distrib. Energy Resour.* **2009**, *5*, 201–226. Available online: https://www.researchgate.net/publication/228344453_Adaptive_network_protection_in_microgrids (accessed on 6 April 2017).
6. Ustun, T.S.; Ozansoy, C.; Zayegh, A. Modeling of a Centralized Microgrid Protection System and Distributed Energy Resources According to IEC 61850-7-420. *IEEE Trans. Power Syst.* **2012**, *27*, 1560–1567. [CrossRef]
7. Zamani, M.A.; Yazdani, A.; Sidhu, T.S. A Communication-Assisted Protection Strategy for Inverter-Based Medium-Voltage Microgrids. *IEEE Trans. Smart Grid* **2012**, *3*, 2088–2099. [CrossRef]
8. Lin, H.; Guerrero, J.M.; Jia, C.; Tan, Z.H.; Vasquez, J.C.; Liu, C. Adaptive overcurrent protection for microgrids in extensive distribution systems. In Proceedings of the IECON 2016, 42nd Annual Conference of the IEEE Industrial Electronics Society, Florence, Italy, 23–26 October 2016; pp. 4042–4047. [CrossRef]
9. Lin, H.; Sun, K.; Tan, Z.; Liu, C.; Guerrero, J.M.; Vasquez, J.C. Adaptive protection combined with machine learning for microgrids. *IET Gener. Transm. Distrib.* **2019**, *13*, 770–779. [CrossRef]
10. Ghadiri, S.M.E.; Mazlumi, K. Adaptive protection scheme for microgrids based on SOM clustering technique. *Appl. Soft Comput.* **2020**, *88*, 106062. [CrossRef]
11. Momesso, A.E.C.; Bernardes, W.M.S.; Asada, E.N. Adaptive directional overcurrent protection considering stability constraint. *Electr. Power Syst. Res.* **2020**, *181*, 106190. [CrossRef]

12. Javadi, M.S.; Nezhad, A.E.; Anvari-Moghaddam, A.; Guerrero, J.M. Optimal Overcurrent Relay Coordination in Presence of Inverter-Based Wind Farms and Electrical Energy Storage Devices. In Proceedings of the 2018 IEEE International Conference on Environment and Electrical Engineering and 2018 IEEE Industrial and Commercial Power Systems Europe (EEEIC/I&CPS Europe), Palermo, Italy, 12–15 June 2018; pp. 1–5. [CrossRef]

13. Samadi, A.; Chabanloo, R.M. Adaptive coordination of overcurrent relays in active distribution networks based on independent change of relays' setting groups. *In. J. Electr. Power Energy Syst.* **2020**, *120*, 106026. [CrossRef]

14. George, S.P.; Ashok, S. Adaptive differential protection for transformers in grid-connected wind farms. *Int. Trans. Electr. Energy Syst.* **2018**, *28*, e2594. [CrossRef]

15. Prasad, C.D.; Biswal, M.; Abdelaziz, A.Y. Adaptive differential protection scheme for wind farm integrated power network. *Electr. Power Syst. Res.* **2020**, *187*, 106452. [CrossRef]

16. Ali, I.; Hussain, S.M.S.; Tak, A.; Ustun, T.S. Communication Modeling for Differential Protection in IEC-61850-Based Substations. *IEEE Trans. Ind. Appl.* **2018**, *54*, 135–142. [CrossRef]

17. Aftab, M.A.; Roostaee, S.; Hussain, S.M.S.; Ali, I.; Thomas, M.S.; Mehfuz, S. Performance evaluation of IEC 61850 GOOSE-based inter-substation communication for accelerated distance protection scheme. *IET Gener. Transm. Distrib.* **2018**, *12*, 4089–4098. [CrossRef]

18. de Sotomayor, A.A.; della Giustina, D.; Massa, G.; Dedè, A.; Ramos, F.; Barbato, A. IEC 61850-based adaptive protection system for the MV distribution smart grid. *Sustain. Energy Grids Netw.* **2018**, *15*, 26–33. [CrossRef]

19. Barra, P.H.A.; Coury, D.V.; Fernandes, R.A.S. A survey on adaptive protection of microgrids and distribution systems with distributed generators. *Renew. Sustain. Energy Rev.* **2020**, *118*, 109524. [CrossRef]

20. Beheshtaein, S.; Cuzner, R.; Savaghebi, M.; Guerrero, J.M. Review on microgrids protection. *IET Gener. Transm. Distrib.* **2019**, *13*, 743–759. [CrossRef]

21. Mahamedi, B.; Fletcher, J.E. Trends in the protection of inverter-based microgrids. *IET Gener. Transm. Distrib.* **2019**, *13*, 4511–4522. [CrossRef]

22. Hussain, N.; Nasir, M.; Vasquez, J.C.; Guerrero, J.M. Recent Developments and Challenges on AC Microgrids Fault Detection and Protection Systems–A Review. *Energies* **2020**, *13*, 2149. [CrossRef]

23. Patnaik, B.; Mishra, M.; Bansal, R.C.; Jena, R.K. AC microgrid protection—A review: Current and future prospective. *Appl. Energy* **2020**, *271*, 115210. [CrossRef]

24. Aftab, M.A.; Hussain, S.M.S.; Ali, I.; Ustun, T.S. IEC 61850 based substation automation system: A survey. *Int. J. Electr. Power Energy Syst.* **2020**, *120*, 106008. [CrossRef]

25. Hatziargyriou, N. *Microgrids: Architectures and Control*; John Wiley & Sons: West Sussex, UK, 2013.

26. Rockefeller, G.D.; Wagner, C.L.; Linders, J.R.; Hicks, K.L.; Rizy, D.T. Adaptive transmission relaying concepts for improved performance. *IEEE Trans. Power Deliv.* **1988**, *3*, 1446–1458. [CrossRef]

27. Relion Protection and Control. *650 Series IEC 61850 Communication Protocol Manual*, Revision: Product version: 1.1; ABB AB Substation Automation Products SE-721 59: Västerås, Sweden, 2011. Available online: https://search.abb.com/library/Download.aspx?DocumentID=1MRK511242-UEN&LanguageCode=en&DocumentPartId=&Action=Launch&DocumentRevisionId=- (accessed on 15 May 2015).

28. Katiraei, F.; Holbach, J.; Chang, T.; Johnson, W.; Wills, D.; Young, B.; Marti, L.; Yan, A.; Baroutis, P.; Thompson, G.; et al. Investigation of Solar PV Inverters Current Contributions during Faults on Distribution and Transmission Systems Interruption Capacity. In Proceedings of the Western Protective Relay Conference, Washington, DC, USA, 16–18 October 2012.

29. Turcotte, D.; Katiraei, F. Fault Contribution of Grid-Connected Inverters. In Proceedings of the 2009 IEEE Electrical Power Conference, Montreal, QC, Canada, 22–23 October 2009.

30. Palizban, O.; Kauhaniemi, K. Energy storage systems in modern grids—Matrix of technologies and applications. *J. Energy Storage* **2016**, *6*, 248–259. [CrossRef]

31. Keller, J.; Kroposki, B. *Understanding Fault Characteristics of Inverter-Based Distributed Energy Resources*; National Renewable Energy Laboratory: Golden, CO, USA, 2010.

32. *Technical Guideline Generating Plants Connected to the Medium-Voltage Network Guideline for Generating Plants' Connection to and Parallel Operation with the Medium-Voltage Network*; German Association of Energy and Water Industries (BDEW): Berlin, Germany, 2008.

33. Hou, D.; Dolezilek, D. IEC 61850-What It Can and Cannot Offer to traditional Protection Schemes, Schweitzer Engineering Laboratories, Inc. *SEL J. Reliab. Power* **2010**, *1*, 20080912.

34. IEC 61850-5:2013. *Communication Networks and Systems in Substations—Part 5: Communication Requirements for Functions and Device Models*; International Electrotechnical Commission, TC 57: Geneva, Switzerland, 2013.

35. Van Rensburg, M.; Dolezilek, D.; Dearien, J. Case Study: Lessons Learned Using IEC 61850 Network Engineering Guideline Test Procedures to Troubleshoot Faulty Ethernet Network Installations, Schweitzer Engineering Laboratories, Inc. In Proceedings of the Power and Energy Automation Conference, Spokane, Washington, DC, USA, 21–23 March 2017. Available online: https://www.eiseverywhere.com/file_uploads/0fca13690b4579341cf53b62d22a0647_Dolezilek1.pdf (accessed on 5 June 2020).

36. IEC/TR 61850-90-1:2010. *Communication Networks and Systems for Power Utility Automation—Part 90-1: Use of IEC 61850 for the Communication between Substations*; International Electrotechnical Commission, TC 57: Geneva, Switzerland, 2010.

37. Aichhorn, A.; Etzlinger, B.; Unterweger, A.; Mayrhofer, R.; Springer, A. Design, implementation, and evaluation of secure communication for line current differential protection systems over packet switched networks. *Int. J. Crit. Infrastruc. Prot.* **2018**, *23*, 68–78. [CrossRef]

38. Chelluri, S.; Dolezilek, D.; Dearien, J.; Kalra, A. Understanding and Validating Ethernet Networks for Mission-Critical Protection, Automation, and Control Applications. March 2014. Available online: https://pdfs.semanticscholar.org/ef57/a6a7655bed807bdd301e7c009d0f2dda4965.pdf (accessed on 2 September 2020).

39. Fred Steinhauser, Lessons Learned Time Related Issues Latency Measurements in Digital Grids, PAC World #Issue 052. June 2020. Available online: https://www.pacw.org/latency-measurements-in-digital-grids (accessed on 28 September 2020).

40. Mekkanen, M.; Kauhaniemi, K.; Kumpulainen, L.; Memon, A.A. Light-Weight IEC 61850 GOOSE Based LOM Protection for Smart Grid. In Proceedings of the CIRED 2018 Ljubljana Workshop on Microgrids and Local Energy Communities, Ljubljana, Slovenia, 7–8 June 2018; p. 0031.

41. Apostolov, A. R-GOOSE: What it is and its application in distribution automation. *CIRED—Open Access Proc. J.* **2017**, *2017*, 1438–1441. [CrossRef]

42. Ustun, T.S.; Khan, R.H.; Hadbah, A.; Kalam, A. An adaptive microgrid protection scheme based on a wide-area smart grid communications network. In Proceedings of the 2013 IEEE Latin-America Conference on Communications, Santiago, Chile, 24–26 November 2013; pp. 1–5.

43. Jeong, Y.W.; Lee, H.W.; Kim, Y.G.; Lee, S.W. High-speed AC circuit breaker and high-speed OCR. In Proceedings of the 22nd International Conference and Exhibition on Electricity Distribution (CIRED 2013), Stockholm, Sweden, 10–13 June 2013; pp. 1–4.

44. Yazdani, A.; Iravani, R. Chapter 9 Controlled-Frequency VSC System. In *Voltage-Sourced Converters in Power Systems: Modeling, Control, and Applications*; John Wiley and Sons, Inc.: Hoboken, NJ, USA, 2010.

45. Kanellos, F.D.; Hatziargyriou, N.D. Control of Variable Speed Wind Turbines in Islanded Mode of Operation. *IEEE Trans. Energy Convers.* **2008**, *23*, 535–543. [CrossRef]

46. Kanellos, F.D.; Hatziargyriou, N.D. Control of variable speed wind turbines equipped with synchronous or doubly fed induction generators supplying islanded power systems. *IET Renew. Power Gener.* **2009**, *3*, 96–108. [CrossRef]

47. SFS-EN 50549-1. Part1: Connection to a LV distribution network. Generating plants up to and including Type B. In *Requirements for Generating Plants to Be Connected in Parallel with Distribution Networks*; European Committee for Standardization: Brussels, Belgium, 2019.

48. Baran, M.; El-Markabi, I. Adaptive over current protection for distribution feeders with distributed generators. In Proceedings of the IEEE PES Power Systems Conference and Exposition, New York, NY, USA, 10–13 October 2004; pp. 715–719.

49. Mahat, P.; Chen, Z.; Bak-Jensen, B.; Bak, C.L. A Simple Adaptive Overcurrent Protection of Distribution Systems with Distributed Generation. *IEEE Trans. Smart Grid* **2011**, *2*, 428–437. [CrossRef]

50. Liu, Z.; Hoidalen, H.K. An adaptive inverse time overcurrent relay model implementation for real time simulation and hardware-in-the-loop testing. In Proceedings of the 13th International Conference on Development in Power System Protection 2016 (DPSP), Edinburgh, UK, 7–10 March 2016. [CrossRef]

51. Liu, Z.; Høidalen, H.K. A simple multi agent system based adaptive relay setting strategy for distribution system with wind generation integration. In Proceedings of the 13th International Conference on Development in Power System Protection 2016 (DPSP), Edinburgh, UK, 7–10 March 2016. [CrossRef]
52. Coffele, F.; Booth, C.; Dyśko, A. An Adaptive Overcurrent Protection Scheme for Distribution Networks. *IEEE Trans. Power Deliv.* **2015**, *30*, 561–568. [CrossRef]

Article

Adaptive Overhead Transmission Lines Auto-Reclosing Based on Hilbert–Huang Transform

Arman Ghaderi Baayeh [1] and Navid Bayati [2],*

[1]   Department of Electrical Engineering, Faculty of Engineering, University of Kurdistan, Sanandaj, Kurdistan 66177-15175, Iran; arman.ghaderi@gmail.com

[2]   Department of Energy Technology, Aalborg University, 6700 Esbjerg, Denmark

*   Correspondence: nab@et.aau.dk

Received: 3 August 2020; Accepted: 14 October 2020; Published: 16 October 2020

Abstract: This paper presents a reliable and fast index to detect the instant of arc extinction for adaptive single-pole automatic reclosing (ASPAR). The proposed method is a simple technique for ASPAR on shunt compensated transmission lines using the Hilbert–Huang Transform (HHT). The HHT method is a combination of the empirical mode decomposition (EMD) and the Hilbert transform (HT). The first intrinsic mode function (IMF1) decomposed by EMD, which contains high frequencies of the faulty phase voltage, was used to calculate the proposed index. HT calculates the first IMF spectrum in the time-frequency domain. The presented index is the sum of all frequency contents below 55 Hz, which remains very low until the fault clearance. The proposed method uses a global threshold level and therefore no adjustment is needed for different transmission systems. This method is effective for various system configurations including different fault locations, line loading, and various shunt reactor configurations, designs, compensation rates, and placement. The performance of the method was verified using 324 test cases simulated in electromagnetic transient program (EMTP) related to a 345 kV transmission line. For all the test cases, the algorithm successfully operated with an average reclosing time delay of 32 ms.

Keywords: adaptive auto-reclosing; power system protection; EV transmission lines; transient fault; Hilbert–Huang transform

---

1. Introduction

In recent years, the protection of transmission lines by reclosing switches has become a challenge of improving the reliability of power systems [1]. Approximately 80% of faults in overhead transmission lines are transient (arcing) and single-phase to earth. As a result, there is no need to permanently de-energize the transmission line and send the repair team to patrol for maintenance purposes, then, actually the fault will be cleared by temporary de-energizing of the transmission line and by reclosing the circuit breakers (CBs), the transmission line can restore to its normal operation. In the case of single-phase faults, isolating the faulty phase is enough and there is no need to three phases reclosing [2]. After fault, the arc current will has an extremely large value and its length is constant, and the fault at this stage is called the primary arc. After the faulty phase is isolated from both sides of the line, the arc is still fed through the healthy phases [3]; the fault at this stage is called the secondary arc. Due to the low current of the secondary arc, the ionized column of the arc becomes narrower and moves with the wind, and its length increases until the extinction of it. The time of the secondary arc that occurred is called the dead-time. Dead-time is much less than the reclose time setting in traditional methods, usually between 0.2–0.8 s, hence single-phase reclosure can be done much faster, which has the following benefits for the power system:

- Improving power system marginal stability during faults;
- Improving in transient stability;
- Improving in system reliability and availability;
- Mitigating of switching over-voltages;
- Mitigating of shaft torsional oscillation of large thermal units;
- Minimizing unsuccessful reclosing; and
- Reducing system and equipment shocks.

Lightning flashover is the prime cause of transient faults. When lightning strikes the tower body or guard wires, the lightning current passes through the tower body and enters the ground. If the grounding resistance is high, the voltage drop across the tower body will also be large, and eventually a flashover will occur between the tower body and the phase conductor. High tower height, high ground resistance, high pollution severity on insulators, and high average isokeraunic level along the transmission line are all factors that increase the likelihood of transient faults occurring in an overhead transmission line. Obviously, it is possible for lightning to strike the tower, but not cause a fault, but this will weaken the insulation properties of the insulators. In general, the high incidence of thunder storms increases the likelihood of transient faults. In ASPAR studies, modeling starts from the primary arc onward, and previous events have no effect on the results of these studies.

Adaptive single phase auto-reclosing methods must be able to quickly and accurately detect the moment of secondary arc extinction. In this regard, various algorithms have been proposed, a few important groups of them are discussed as follows.

Due to the quasi-square waveform of the arc voltage, the faulty phase voltage contains the odd harmonics of the fundamental component. After the fault clearance, the values of these harmonics decrease and will ideally reach zero. The approaches presented in [4–10] calculated the harmonic content of the faulty phase voltage or healthy phase currents using various signal processing techniques including time-time (TT) transform, discrete wavelet transform, wavelet packet, and total harmonic distortion (THD). Finally, the fault nature and the moment of extinction of the secondary arc were detected using the calculated criterion values and changings. In the presence of renewable resources, the value of THD is always greater than zero, and this disrupts the performance of such methods.

In [11], the third harmonic of the zero sequence voltage at the local substation was used as a criterion to detect secondary arc extinction. Using voltage measurements at both sides of the line was suggested in [12] by a communication-aided index for ASPAR based on predicted and measured voltage of the faulty phase. The approach proposed in [13] computes the secondary arc current based on measurements to decrease reclosing delay. The proposed real-time method requires signals measured on both sides of the transmission line, however, it can continue to work with local measurements.

In [14], based on local faulty phase voltage and the adaptive cumulative sum method, an increase in voltage amplitude due to fault clearing was recognized. The algorithm [15] utilized the least square method to predict present voltage magnitude value; at the arc extinction moment, the difference between predicted and measured voltage magnitude increased and an adaptive threshold-less approach detected fault extinction. Ghaderi-Baayeh, in [16], introduced a new method for ASPAR based on the second derivative of the faulty phase local voltage angle to determine the secondary arc extinction time in transient fault cases. In [17], the absolute value of the first derivative of the faulted phase local voltage measurement was used to detect the secondary arc extinction. The proposed algorithm has fast performance and uses a low sampling frequency rate and adaptive threshold value. In [18], based on local voltage measurements, a combination of voltage and angle first derivation was utilized to identify fault type and arc extinction detection.

In [19], based on traveling wave theory and using local measured voltage for three types of mixed transmission systems, the occurrence of fault and its location are determined. In the case of fault in the overhead line section, reclosure permission was issued. In [20], for mixed transmission systems, based on wavelet transform and the difference between the currents in the active part of the cable and those in the shields, the overhead section fault was detected. Reclosure into a permanent fault caused damage to the generator shaft of nearby power plants. Since the reclose commands issued for each side of the line were not synchronous, one side of the line always reclosed faster. In [21], a method for selecting the side of the line that should lead reclosing was proposed. In [22], the fault nature, whether permanent or transient, was determined based on the locally measured voltage and using a featured classifier based on support vector machine.

Shunt reactors are widely used in high voltage transmission lines to improve power system stability and line voltage profile regulation [23]. The methods presented in [24–30] are effective for shunt compensation transmission lines. In these methods, the beat frequency generated after the quenching of the secondary arc was used to detect the quenching of the secondary arc. In [24], the local measured voltage frequency was analyzed using modal transformation and a simple zero crossing algorithm. The secondary arc extinction time was then specified for single- and double-phase-to-ground faults. In [25], mode currents of shunt reactors were calculated, then the presence and number of natural frequencies were used to distinguish the fault nature. In [26], the differences of faulty phase terminal voltage between the two fault states after arc extinction was utilized as a criterion to detect fault nature and clearance. In [27,28], the instantaneous power algorithm was utilized to compute faulty phase reactive and active power, respectively, using local voltage and current measurements. Increasing faulty phase reactive or active power after secondary arc extinction was used to detect arc extinction for shunt compensated transmission lines. In [29], based on the cascaded delayed signal cancellation technique and using the faulted phase local voltage of the shunt compensated line, the average distortion rate was calculated to identify the fault nature and clearance.

The majority of ASPARs presented so far only used the measured data on one side of the transmission line, but references [12,13,30] required measured data on both sides of the transmission line. In [30], the presence of phasor measurement units (PMUs) on both sides of the transmission line was necessary to identify the type of fault and its clearance time.

Long and high voltage transmission lines are very prone to transient faults. On the other hand, to prevent overvoltage, most of these lines are shunt compensated on either side or one side. After the secondary arc extinction, the trapped energy oscillation between the line capacitor and the reactor inductance creates a sub-synchronous component in the isolated phase voltage. In this study, by using the presence of sub-synchronous components in the faulty phase voltage spectrum, an index was proposed to detect fault clearance. During the fault, the voltage of the faulty phase does not contain any sub-synchronous components, however, after extinguishing of the secondary arc, due to the resonance between the shunt reactor and the line capacitor, a sub-synchronous component appears in the voltage. This change in frequency content from 0 to 60 Hz is introduced to detect secondary arc extinction. The Hilbert-Huang Transform (HHT) method is used in this algorithm, which can ideally monitor sub-synchronous components with very low spectrum leakage and high accuracy. Since the proposed method, unlike many previous methods, does not use the THD of measured signals, consequently, it is not sensitive to the presence of renewable resources and is a good option for protecting the grid in the high penetration of renewable resources. System simulation studies show that the proposed algorithm estimates the fault clearing instant accurately for auto-reclosing.

The rest of the paper is structured as follows. Section 2 describes the modeling of the understudy system and Section 3 provides an introduction to the Hilbert–Huang transform. Section 4 presents the new algorithm followed by the simulation results in Section 5 and conclusions.

2. Modeling of Understudy System

Figure 1 shows the understudy 345 kV transmission line with the length of 200 km. The network had two buses that were connected to the power grid via two Thevenin equivalent voltage sources. The short circuit levels of the U1 and U2 equivalents were 10 GVA and 15 GVA, respectively. Additionally, the x/r ratio of the positive and zero sequences was 10 and 4, respectively. Table 1 shows the parameters of the single circuit transmission line for the positive and zero sequences. This transmission line was transposed and modeled as the frequency-dependent line in the EMTP software environment. The shunt reactor, as shown in Figure 1, had four windings and was grounded through an inductance to mitigate secondary arc current. The shunt reactor parameter calculations are expressed as follows for different cases.

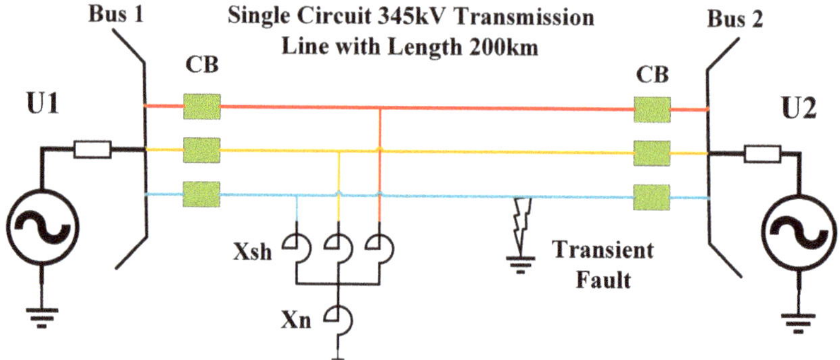

Figure 1. The under study 2-bus power system.

Table 1. The 345 kV transmission line with 200 km length.

	R (ohm)	L (H)	C (uF)
Zero sequence	2.8813	0.3647	1.4713
Positive sequence	0.0515	3.1572	2.3269

Secondary Modeling

In this paper, the Kizilcay method for secondary arc modeling in EMTP software was used. More details can be found in [31], which are not repeated here for brevity. The specifications of the arc used in the simulations were as follows:

Initial resistance (r_0): 150 mΩ

Voltage per unit of length (u_0): 1.1 V/cm

Initial time constant (τ_0): 960 ms

Initial length (l_0) of arc: 3.5 m

3. Hilbert–Huang Transform

An analytic signal is a complex-valued function without any negative frequency components. The real and imaginary parts of an analytic signal are real-valued functions related to each other by the Hilbert transform (HT). The HT of the real function $r(t)$ is equal to $\hat{r}(t)$. The definition of HT is as follows:

$$\hat{r}(t) = \frac{1}{\pi} P \int_{-\infty}^{+\infty} \frac{r(\tau)}{t - \tau} d\tau = r(t) * \frac{1}{\pi t} \tag{1}$$

where the symbol '*' denotes the convolution operation and P indicates the Cauchy principal value. The analytic function of $r(t)$ with respect to time can be defined as follows:

$$\overline{r}(t) = r(t) + j\hat{r}(t) = m(t)e^{i\varphi(t)} \tag{2}$$

$$m(t) = \sqrt{r^2 + \hat{r}^2}, \quad \varphi(t) = \tan^{-1}\frac{\hat{r}}{r} \tag{3}$$

where $j = \sqrt{-1}$ and $\bar{\bar{r}}(t)$ is the analytic function calculated for the real function $r(t)$. m and φ are the instantaneous energy and phase functions in terms of time, respectively. It is unnecessary to compute the integral in (1) to achieve the HT of $r(t)$. Instead, in the first step, the Fourier transform of the real function $r(t)$ is calculated, then its negative frequency components are set to zero and the inverse Fourier transform applied to it. The obtained function is the analytic function $\bar{\bar{r}}(t)$ calculated for the real function $r(t)$, which its imaginary part is $\hat{r}(t)$. The instantaneous frequency of the real function $r(t)$ is:

$$f(t) = \frac{1}{2\pi}\frac{d\varphi}{dt} \tag{4}$$

The method of calculating the frequency-time distribution for the square amplitude of $r(t)$ can be explained using HT. HT is a very suitable method for single mode signals and also calculates the energy distribution in time and frequency. However, the majority of signals are multi-components and their frequency content is spread in the frequency domain. In HHT method, to solve this problem, the input signal was first divided into several intrinsic mode functions (IMFs) by the empirical mode decomposition (EMD) method, where IMFs are time-varying mono-component (single frequency) functions [32]. Then, each of these IMFs is transformed by HT to the frequency domain and their energy is expressed in terms of time and frequency. Using the sifting process, the input signal $S(t)$ is decomposed in terms of IMFs [32]:

$$S(t) = \sum_{i=1}^{n} IMF_i(t) + R(t) \tag{5}$$

where $R(t)$ is the residual function and n is the number of IMFs. IMF_i is the ith decomposed IMF. The analytic functions of IMFs are calculated using the method described above. Each of these functions $\overline{\overline{IMF}}_i(f,t)$ represents part of the original signal spectrum. $f_i(t)$ and $m_i(t)$ vectors are calculated for each IMF_i in the time domain. Finally, the 2n vector, with the same length as the time vector t, are the output of the HHT. The spectrum of the input signal $\bar{\bar{S}}(f,\ t)$ is the combination of these vectors:

$$\bar{\bar{S}}(f,\ t) = \sum_{i=1}^{n} sparse(t,\ f_i(t),\ m_i(t)) = \sum_{i=1}^{n} \bar{\bar{S}}_i(f,\ t) \tag{6}$$

where m_i and f_i are the instantaneous energy and frequency functions in terms of time for IMF_i, respectively. Moreover, sparse matrix is a matrix that has a large number of zeros. *sparse* function generates a sparse matrix $\bar{\bar{S}}_i(f,\ t)$ from the triplets t, $f_i(t)$, and $m_i(t)$ such that $\bar{\bar{S}}_i(f_i(k),\ t(k)) = m_i(k)$. Entries that have no value assigned to them are equal to zero. The lengths of the three vectors t, $f_i(t)$, and $m_i(t)$ are equal.

4. Proposed Method

Many high voltage transmission lines have shunt compensation, which is placed on either side of the transmission line or one side only. Reactors are usually grounded by inductance to reduce the flow of secondary arc current, and after the fault clearance, the shunt reactor and capacitor of the transmission line are parallel to each other. The energy trapped in the reactor and shunt capacitor of the transmission line oscillates until complete damping, and this energy oscillation between the transmission line and the parallel compensation creates a sub-synchronous component in the isolated phase voltage. In the literature, the created resonance after fault clearance has been used to diagnose secondary arc extinction [15,18,24–29]. There is no sub-frequency component during fault, however, after the secondary arc extinguishing in the frequency range of 0 to 60 Hz, a sub-synchronous component appears in the faulty phase voltage for compensated transmission lines. Therefore, in this paper, the energy in the bandwidth of 0 to 55 Hz of the faulty phase voltage was used as a criterion for detecting secondary arc extinction. The frequency of these sub-synchronous components is practically between 30 Hz and 45 Hz for 60 Hz power systems [33].

Due to the limitations on the computational burden, a small data window should be used in protection system, and on the other hand, there is an edge effect in all signal processing methods, hence, both of these constraints should always be considered in choosing the length of the data window. The following index is proposed as a criterion for detecting secondary arc extinction:

$$IN(i) = \sum_{f=0}^{55} \bar{\bar{U}}_1\left(f, \frac{WL}{2}\right) \tag{7}$$

where IN represents the sum of the energy of all sub-synchronous components in the faulty phase voltage. U is the last measured window of the faulty phase voltage with the length of WL. U_1 is the first decomposed IMF of U. $\bar{\bar{U}}_1$ is the output of the Hilbert transformation of U_1, where the central sample of the data window was used here to calculate IN for avoiding the edge effect. HHT can extract sub-synchronous components without being affected by the fundamental 60 Hz component.

Figure 2a shows the faulty phase voltage at Bus 1 for the single line to ground transient fault at 30% of the line. The reactance of the effectively grounded shunt compensator at Bus 2 was 1506.60 Ω. The fault occurred at 350 ms and after 150 ms, the faulty phase was completely isolated from both sides. From 350 to 500 ms, the transient fault had an extremely large current without any differences in its characteristics from the permanent fault. Transient fault at this stage is called the primary arc and its length is almost constant. From the moment of 0.5 s onward, the arc length increased slowly until the transient fault was cleared. The simulation was performed for 1.2 s. Figure 2b shows the spectrum extracted by HHT. The HHT method has a small leakage spectrum and the different components have little effect on each other. Figure 2c illustrates the proposed index for a 1920 Hz sampling frequency (32 samples per 60 Hz cycle) and WL = 50 ms, where the value is zero during the secondary arc and increases after the secondary arc is extinguished.

Figure 3 shows the first three extracted IMFs from the voltage waveform in Figure 2a. The voltage waveform was decomposed to 10 IMFs, but high-order IMFs had very little energy and are not shown here. In the EMD method, IMF_1 always contains high-frequency contents of the input signal. As shown in Figure 3a,b, the sub-synchronous component only appeared in the IMF_1 of the frequency spectrum. Thereby, IMF_1 is the most suitable IMF to monitor the sub-synchronous component due to secondary arc extinction.

Figure 4 shows the flowchart of the proposed algorithm. The proposed method detects fault clearance for shunt compensated transmission lines, and recognition of permanent or transient faults was not within the scope of this paper, and the fault type was assumed to be a transient single line to ground (SLG) fault. After the single-pole operation of the CBs on both sides of the transmission line, the faulty phase was isolated. It was assumed that the distance protections on both sides of the transmission line quickly de-energized the faulty phase. Some papers have assumed that the fault was transient and waited for the secondary arc extinction. In this case, if the secondary arc is not detected after a certain time (about 1.5 to 3 s), it is concluded that the fault is not transient and the initial assumption is wrong, and the three-phase trip command is issued. In this case, the secondary arc extinction detection algorithm is used to detect the nature of the fault. The problem with this fault nature recognition is the delay of about 1.5 to 3 s, which means that the system has been in two phases for this period in the presence of a permanent fault for no reason, and this causes the power system to move more toward instability. In another category of papers, a separate method was proposed to identify the nature of the fault, which did this much faster than the first case. These methods usually use the presence of odd harmonics during the secondary arc or the presence of high frequency components during the primary arc at voltage. In any case, the detection of fault instant, fault location, and fault nature was not in the scope of this paper and the contribution of this paper is in the detection of secondary arc extinction.

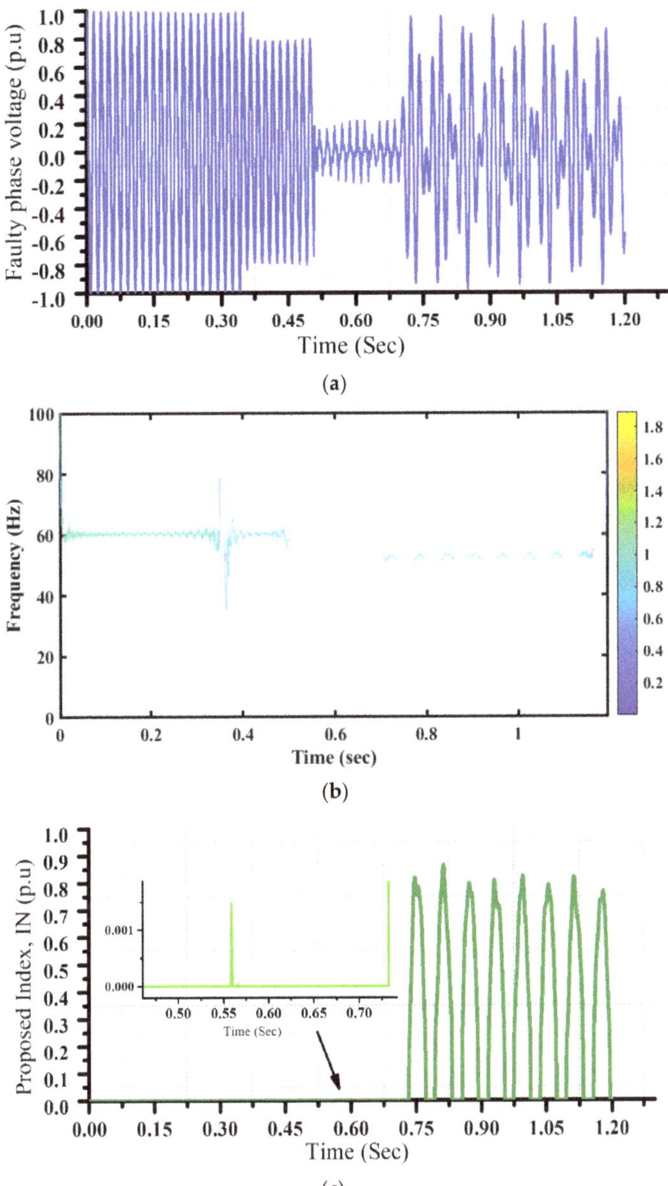

Figure 2. The faulty phase voltage at Bus 1 for the single line to ground transient fault. (**a**) The faulty phase voltage U (p.u.). (**b**) The spectrum extracted by Hilbert-Huang Transform (HHT), and (**c**) the proposed index IN.

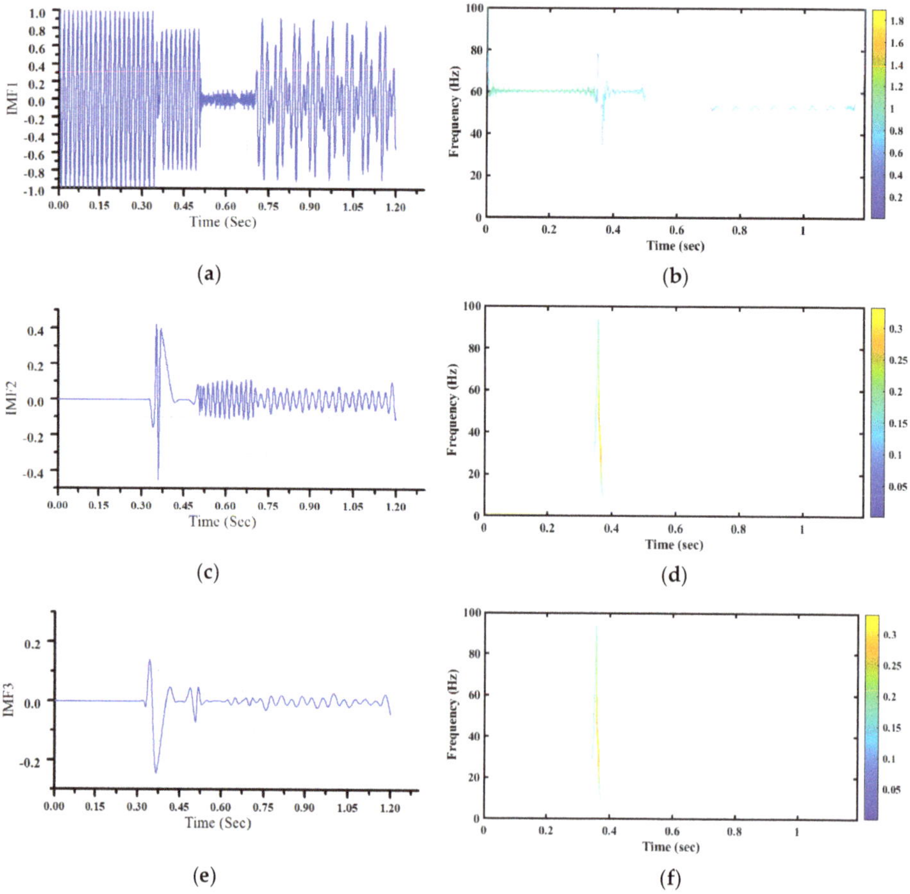

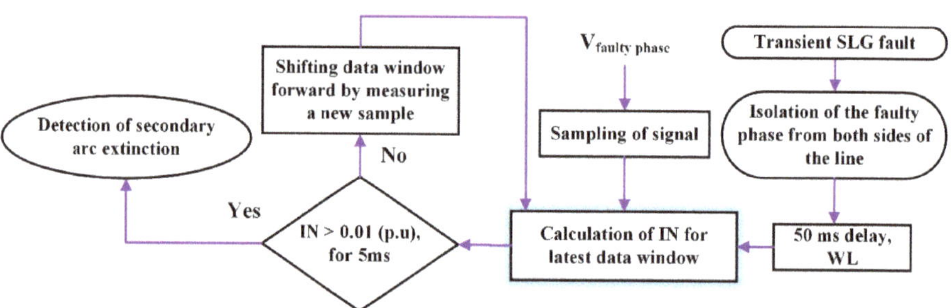

Figure 3. Intrinsic mode functions (IMFs) extracted from the voltage waveform in Figure 2a and their extracted spectrum using the HT (**a**) IMF_1 (**b**) IMF_1 spectrum (**c**) IMF_2 (**d**) IMF_2 spectrum (**e**) IMF_3 (**d**) IMF_3 spectrum.

Figure 4. Flowchart of the proposed algorithm.

The algorithm was delayed 50 ms, thus all samples in the data window were measured after faulty phase isolation, next, the algorithm began to calculate IN using the faulty phase voltage. In the simulations performed in this paper, IN, during the secondary arc, was in most cases zero and sometimes takes very small values. However, after the secondary arc extinction, its value increases and must be greater than the threshold value of 0.01 p.u. for 5 ms to confirm fault clearance. The counting strategy is used in digital protection to improve the reliability of the relay operation or avoid relay maloperation. Checking a protection decision for several times improves the probability of successful operation.

Due to the presence of noise in real signals and to prevent protecting system mal-operation, IN was not compared with zero or smaller threshold value, however, the proposed criterion did not need any pre-calculation of the threshold value and the threshold value had a global feature. After each comparison of IN with the threshold level, the data window shifted forward and the new IN value was calculated. After secondary arc extinction and deionization of the arc path, the reclose command was issued for the local CB.

The main contributions of this paper include the following:

- Extraction and monitoring of sub-synchronous voltage component using HHT, which is a powerful and suitable tool for this application.
- Proposing a criterion that is zero in most moments and cases during the secondary arc and takes a large amount immediately after the secondary arc is extinguished.
- The proposed criterion is very insensitive to voltage magnitude oscillations during the secondary arc, unlike many existing methods.

5. Simulations and Results

To verify and validate the effectiveness and accuracy of the proposed method for ASPAR, various simulation tests were carried out under different fault locations, line loadings, and various shunt reactor configurations, designs, compensation rates, and placements. The EMTP software was utilized with a 50 ms data window length (WL) and 1.92 kHz sampling frequency rate (32 samples per cycle for 60 Hz).

The reactors used in the simulations of this paper were grounded efficiently or to reduce the secondary arc current, and the three-phase shunt reactors were grounded through a reactance X_n. In the literature, two methods for calculating X_n have been proposed [34–36]. The method presented in [34,35] is independent of the transmission line length, but reduced the fault current less than the method presented in [36]. The reactance value calculated according to the method in [34,35] was always less than the [35] method. The following is the calculation of X_n based on these two methods.

$$F = \frac{1}{X_{sh} \times B_1} \tag{8}$$

$$X_n = \frac{X_{sh}}{3} \times \left(\frac{B_1}{B_0} - 1 \right) (\Omega) \tag{9}$$

$$X_n = \frac{B_1 - B_0}{3F \times B_1 (B_0 - (1 - F) B_1)} (\Omega) \tag{10}$$

where B_0, B_0, F, X_{sh}, and X_n are positive sequence line susceptance, zero sequence line susceptance, shunt compensation rate, the equivalent reactance of line shunt reactor, and equivalent reactance of neutral reactor, respectively. The reactances calculated in (9) and (10) corresponded to the placement of the reactor only on one side of the line. If the transmission line is compensated from both sides, the reactance value of each reactor on each side of the line was twice of the X_{sh} and X_n reactances calculated above. In other words, the reactive power consumption of the equivalent reactor was divided into two. Table 2 shows the number of shunt reactors for the placement of one or both sides of the line, three rates of compensation 0.5, 0.75, and 0.95, and three modes of effective grounding and grounding based on the methods in [34–36].

Table 2. Reactance values of the shunt reactors for the simulations.

Shunt Compensation Percentage		50%		75%		95%	
	Reactance (ohm)	X_{sh}	X_n	X_{sh}	X_n	X_{sh}	X_n
Effectively grounded	Reactor at only one side	2245.80	-	1506.60	-	1190.30	-
	Reactor at both ends	4491.60	-	3013.20	-	2380.60	-
Grounded through neutral reactor, calculated based on [36]	Reactor at only one side	2245.80	1967.10	1506.60	474.77	1190.30	247.23
	Reactor at both ends	4491.60	3934.20	3013.20	949.55	2380.60	494.45
Grounded through neutral reactor, calculated based on [34,35]	Reactor at only one side	2245.80	435.30	1506.60	292.04	1190.30	230.71
	Reactor at both ends	4491.60	870.61	3013.20	584.08	2380.60	461.42

Table 3 shows the performance of the proposed technique for different power system conditions and types of shunt compensation. Fault clearing detection delay (FCDD) is defined as the time difference between the secondary arc extinction detection and the secondary arc extinction moment obtained from the simulation. Simulations were performed for medium and high loading of the transmission line and various fault locations. By measuring the faulty phase voltage on both sides of the transmission line, the algorithm was tested for both substations adjacent to the line. As can be seen, regardless of the power system conditions and the type and rate of compensation, the proposed scheme can accurately and quickly detect fault clearance. The average FCDD for 324 tested cases was 32 ms.

Figure 5a shows the faulty phase voltage at Bus 2 for the single line to ground fault at 30% of the line. The X_{sh} and X_n of the inductively grounded shunt compensators at each side of the lines were 2380.60 and 461.42 Ω, respectively. Fundamental voltage magnitude fluctuations cause energy spreadation in low frequency components on the most signal processing methods. The HHT method adaptively distinguishes between fundamental voltage magnitude fluctuations and the presence of a low frequency component, therefore the spectrum extracted from the sub-synchronous components receives the least effect from the fundamental component. During the secondary arc, due to the reduction of the fault current, the length of the arc fluctuates more, and as a result, the voltage of the faulty phase fluctuates severely (Figure 5a). As demonstrated in Figure 5b, the proposed criterion is independent of the arc behavior and the IN value is zero during the secondary arc. This is due to the use of IMF_1 and the excellent ability of HHT to decompose signal components with minimal spectrum leakage. Similar waveforms of field measurements, obtained from electrical systems reported in [37], showed similar voltage behavior, hence, it is important that the secondary arc extinction detection method is independent of the voltage behavior during the dead-time.

Due to the growth of renewable resource penetrations in the power system, the need for adaptive protection methods is increasing. Renewable resources are generally connected to the power system through power electronic interfaces, and one of their destructive effects is increasing the power system THD [38]. Many of the methods proposed in the literature are based on the increase of THD during the secondary arc and the sharp decrease after the fault clearance. In the presence of renewable resources, many of these methods lose their ability to function properly due to the lack of THD reduction after fault clearance [4–10]. The proposed method does not use the harmonics in voltage to calculate the criterion, and in addition, due to the high capability of the HHT method, the proposed method can have a correct and fast performance regardless of the high penetration of renewable resources in the power system and THD value.

Table 3. Fault clearing detection delay (FCDD, ms) for various power system and fault conditions and in the presence of the shunt reactor.

FCDD for ASPAR Relay at Substation1/Substation2			Medium Loading, 50% of the Line SIL			High Loading, 100% of the Line SIL		
fault location from Bus1			30%	60%	90%	30%	60%	90%
Effectively grounded	Reactor at Bus1	50%	32/34	29/34	34/30	35/32	32/36	31/31
		75%	35/32	32/31	30/32	33/30	30/34	32/29
		95%	28/32	31/33	28/29	31/31	32/34	33/32
	Reactor at Bus2	50%	32/34	31/32	29/32	33/32	32/30	33/34
		75%	30/33	30/34	33/31	32/35	33/33	31/32
		95%	38/33	31/33	33/36	30/30	32/31	32/31
	Reactor at both ends	50%	34/35	28/34	33/34	34/34	33/31	36/32
		75%	35/34	34/32	32/27	32/33	33/31	28/30
		95%	30/33	34/32	33/33	32/32	30/30	37/31
Grounded through neutral reactor, calculated based on [36]	Reactor at Bus1	50%	31/31	32/34	31/29	33/30	31/30	33/32
		75%	32/31	32/28	34/32	32/32	31/32	34/32
		95%	34/35	31/31	29/32	30/32	32/29	29/32
	Reactor at Bus2	50%	32/29	35/30	33/34	31/29	32/33	31/33
		75%	34/32	32/31	30/31	31/33	33/32	32/32
		95%	28/28	31/34	31/35	31/31	26/33	33/36
	Reactor at both ends	50%	31/34	35/34	33/31	35/32	32/32	29/33
		75%	30/34	32/30	34/33	30/33	35/33	33/36
		95%	29/32	31/31	30/33	31/30	30/29	30/31
Grounded through neutral reactor, calculated based on [34,35]	Reactor at Bus1	50%	33/32	32/32	35/32	30/32	34/30	32/33
		75%	33/27	31/32	33/34	29/33	33/31	33/32
		95%	32/33	30/33	32/33	31/29	32/32	33/33
	Reactor at Bus2	50%	29/28	38/33	32/30	31/34	33/34	34/33
		75%	34/28	36/30	32/28	35/37	29/31	34/28
		95%	33/32	33/32	32/36	31/34	32/34	31/29
	Reactor at both ends	50%	32/30	30/28	32/31	30/32	36/33	33/29
		75%	32/33	31/34	32/32	35/33	33/34	30/33
		95%	32/34	32/31	34/33	34/30	33/32	29/35

In Table 4, a group of recent papers have been selected for qualitative comparison with the presented method. As can be seen, the strengths of the proposed method include the following:

- Low starting time delay;
- Fast reclosing time delay (with average 32 ms);
- Medium sampling frequency (1.920 kHz);
- Very low sensitivity to voltage magnitude oscillation; and
- Global threshold.

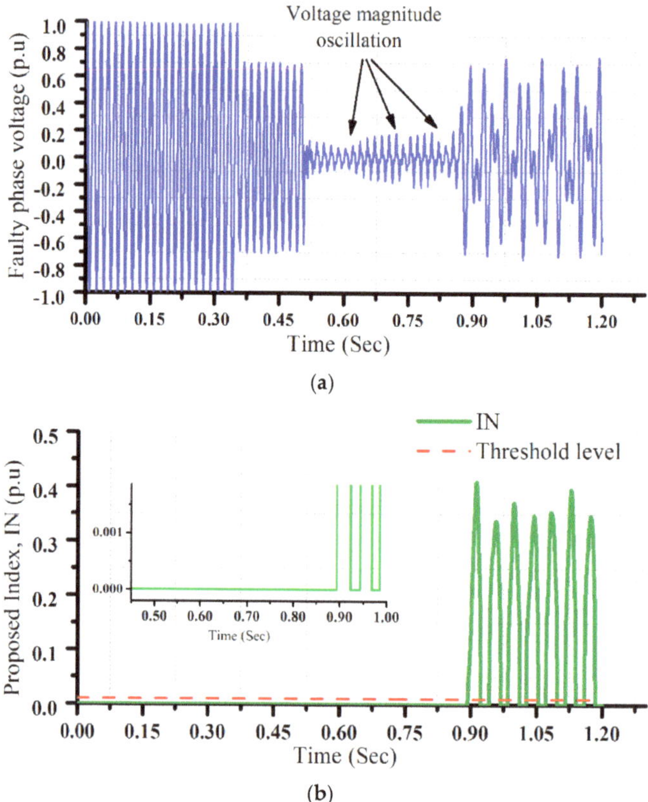

Figure 5. Faulty phase voltage oscillation during the secondary arc. (**a**) Faulted phase voltage waveform (p.u.). (**b**) Proposed index, IN (p.u.).

Table 4. Comparison of the methods for detection of secondary arc extinction.

Method	Starting Time Delay, (ms)	Reclosing Time Delay (ms)	Communication-Based?	Sampling Frequency (Hz)	Threshold	Computation Burden	Sensitivity to Voltage Magnitude Oscillation
[4]	80	<15	No	3200	Adaptive	Medium	Low
[5]	200	<76	No	1920	Fixed	Medium	Low
[8]	60	<60	No	2500	Fixed	High	Low
[10]	120	<10	No	5000	Not used	Low	Very High
[11]	60	<9	No	1600	Not used	Low	Low
[14]	100	<4	No	1000	Fixed	Low	Very High
[16]	Adaptive, with average of 31 ms	<31	No	800	Adaptive	Low	Medium
[18]	NM	<105	No	3840	Fixed	Medium	Medium
[12]	NM	<67	Yes	3840	Fixed	Medium	low
[30]	40	<14	Yes	720	Fixed	Medium	low
Proposed method	50 ms	<38	No	1920	Global	Low	Very low

NM = Not mentioned.

6. Conclusions

In this paper, a novel algorithm for detecting secondary arc extinction in shunt compensated transmission lines was presented. Using the Hilbert–Huang transform (HHT), the absolute value of all frequency contents below 60 Hz of the fault phase was calculated. Secondary arc extinction was then detected using a global threshold value. The algorithm was tested on a transmission line for different fault location, line loading, and various shunt reactor configurations, designs, compensation rates, and placement. The test results validate the algorithm's performance as well as its independence of the shunt compensation configuration. The main features of the algorithm can be summarized as follows:

- No adjustment is needed for different transmission systems, threshold is global.
- Fast operation time.
- Non-communication based (i.e., use of local measurements).
- Independent from voltage magnitude variation during dead-time.
- Using voltage phasor where it is more reliable rather than the current.
- Fast operation time; average reclosing time delay is 32 ms.
- Applicable for high penetration of renewable resources in the power system.
- Applicable for all types of shunt reactor configurations.
- Algorithm starting time delay is 50 ms, as a result, the algorithm can work properly. even when the fault clears very fast.
- Using the HHT, which is very powerful and practical in the analysis of nonstationary and nonlinear signals.

Author Contributions: Methodology, validation, writing—original draft preparation, A.G.B.; Review and editing, supervision, N.B. All authors have read and agreed to the published version of the manuscript.

Funding: This research received no external funding.

Conflicts of Interest: The authors declare no conflict of interest.

References

1. Bayati, N.; Hajizadeh, A.; Soltani, M. Protection in DC microgrids: A comparative review. *IET Smart Grid* **2018**, *1*, 66–75. [CrossRef]
2. Bayati, N.; Baghaee, H.R.; Hajizadeh, A.; Soltani, M. A Fuse Saving Scheme for DC Microgrids with High Penetration of Renewable Energy Resources. *IEEE Access* **2020**, *8*, 137407–137417. [CrossRef]
3. Bayati, N.; Baghaee, H.R.; Hajizadeh, A.; Soltani, M. Localized protection of radial DC microgrids with high penetration of constant power loads. *IEEE Syst. J.* **2020**. [CrossRef]
4. Nikoofekr, I.; Sadeh, J. Determining secondary arc extinction time for single-pole auto-reclosing based on harmonic signatures. *Electr. Power Syst. Res.* **2018**, *163*, 211–225. [CrossRef]
5. Dias, O.; Tavares, M.C.; Magrin, F. Hardware implementation and performance evaluation of the fast adaptive single-phase auto reclosing algorithm. *Electr. Power Syst. Res.* **2019**, *168*, 169–183. [CrossRef]
6. Jamali, S.; Ghaffarzadeh, N. Adaptive single-pole auto-reclosure for transmission lines using sound phases currents and wavelet packet transform. *Electr. Power Compon. Syst.* **2010**, *38*, 1558–1576. [CrossRef]
7. Jannati, M.; Moshtagh, J. Hardware implementation of a real-time adaptive single-phase auto-reclosure for power transmission lines. *Int. Trans. Electr. Energy Syst.* **2020**, *30*, e12344. [CrossRef]
8. Jamali, S.; Ghaffarzadeh, N. Adaptive single pole auto-reclosing using discrete wavelet transform. *Eur. Trans. Electr. Power* **2011**, *21*, 973–986. [CrossRef]
9. Luo, X.; Huang, C.; Jiang, Y.; Guo, S. An adaptive three-phase reclosure scheme for shunt reactor-compensated transmission lines based on the change of current spectrum. *Electr. Power Syst. Res.* **2018**, *158*, 184–194. [CrossRef]
10. Jannati, M.; Nazari, M.Y. Improving the stability of power transmission lines based on an adaptive single pole auto-reclosure using a two-step strategy. *IET Gener. Transm. Distrib.* **2020**, *14*, 873–882. [CrossRef]

11. Jamali, S.; Parham, A. New approach to adaptive single pole auto-reclosing of power transmission lines. *IET Gener. Transm. Distrib.* **2010**, *4*, 115–122. [CrossRef]

12. Zadeh, M.D.; Rubeena, R. Communication-aided high-speed adaptive single-phase reclosing. *IEEE Trans. Power Deliv.* **2012**, *28*, 499–506. [CrossRef]

13. Vogelsang, J.; Romeis, C.; Jaeger, J. Real-time adaption of dead time for single-phase autoreclosing. *IEEE Trans. Power Deliv.* **2015**, *31*, 1882–1890. [CrossRef]

14. Khodadadi, M.; Noori, M.R.; Shahrtash, S.M. A noncommunication adaptive single-pole autoreclosure scheme based on the ACUSUM algorithm. *IEEE Trans. Power Deliv.* **2013**, *28*, 2526–2533. [CrossRef]

15. Jamali, S.; Baayeh, A.G. Detection of secondary arc extinction for adaptive single phase auto-reclosing based on local voltage behaviour. *IET Gener. Transm. Distrib.* **2017**, *11*, 952–958. [CrossRef]

16. Baayeh, A.G.; Jamali, S. Detection of Secondary Arc Extinction for Adaptive Single Phase Auto-Reclosing. In Proceedings of the 11th International Conference on Protection and Automation in Power System IPAPC, Tehran, Iran, 17–18 January 2017.

17. Baayeh, A.G.; Jamali, S. Adaptive Single Phase Auto-Reclosing Using Voltage Magnitude Pattern. In Proceedings of the 31th International Power System Conference, Tehran, Iran, 24–26 October 2016.

18. Zhalefar, F.; Zadeh, M.R.; Sidhu, T.S. A high-speed adaptive single-phase reclosing technique based on local voltage phasors. *IEEE Trans. Power Deliv.* **2015**, *32*, 1203–1211. [CrossRef]

19. Sohn, S.H.; Cho, G.J.; Kim, C.H. A Study on Application of Recloser Operation Algorithm for Mixed Transmission System Based on Travelling Wave Method. *Energies* **2020**, *13*, 2610. [CrossRef]

20. Granizo Arrabe, R.; Platero Gaona, C.A.; Alvarez Gomez, F.; Rebollo Lopez, E. Novel Auto-Reclosing Blocking Method for Combined Overhead-Cable Lines in Power Networks. *Energies* **2016**, *9*, 964. [CrossRef]

21. Cho, G.J.; Park, J.K.; Sohn, S.H.; Chung, S.J.; Gwon, G.H.; Oh, Y.S.; Kim, C.H. Development of a Leader-End Reclosing Algorithm Considering Turbine-Generator Shaft Torque. *Energies* **2017**, *10*, 622.

22. Morales, J.; Muñoz, E.; Orduña, E.; Idarraga-Ospina, G. A novel approach to arcing faults characterization using multivariable analysis and support vector machine. *Energies* **2019**, *12*, 2126. [CrossRef]

23. Khan, W.A.; Bi, T.; Jia, K. A review of single phase adaptive auto-reclosing schemes for EHV transmission lines. *Prot. Control Mod. Power Syst.* **2019**, *4*, 18. [CrossRef]

24. Dantas, K.M.; Neves, W.L.; Fernandes, D. An approach for controlled reclosing of shunt-compensated transmission lines. *IEEE Trans. Power Deliv.* **2013**, *29*, 1203–1211. [CrossRef]

25. Huang, X.; Song, G.; Wang, T.; Gu, Y. Three-phase adaptive reclosure for transmission lines with shunt reactors using mode current oscillation frequencies. *J. Eng.* **2018**, *25*, 1012–1017. [CrossRef]

26. Jiaxing, N.; Baina, H.; Zhenzhen, W.; Jie, K. Algorithm for adaptive single-phase reclosure on shunt-reactor compensated extra high voltage transmission lines considering beat frequency oscillation. *IET Gener. Transm. Distrib.* **2018**, *12*, 3193–3200. [CrossRef]

27. Xie, C.; Li, F.; Wang, B.; Fan, Y.; Jing, L.; Chen, W.; Wang, C. Anti-interference adaptive single-phase auto-reclosing schemes based on reactive power characteristics for transmission lines with shunt reactors. *Electr. Power Syst. Res.* **2019**, *170*, 176–183. [CrossRef]

28. Xie, C.; Li, F.; Fan, Y.; Wang, C.; Wang, T. Faulty phase active power characteristics-based adaptive single-phase reclosing schemes for shunt reactors-compensated wind power outgoing line. *Wind Energy* **2019**, *22*, 1746–1759. [CrossRef]

29. Xie, X.; Huang, C. A novel adaptive auto-reclosing scheme for transmission lines with shunt reactors. *Electr. Power Syst. Res.* **2019**, *171*, 47–53. [CrossRef]

30. Khorashadi-Zadeh, H.; Li, Z. Design of a novel phasor measurement unit-based transmission line auto reclosing scheme. *IET Gener. Transm. Distrib.* **2011**, *5*, 806–813. [CrossRef]

31. Kizilcay, M.; Pniok, T. Digital simulation of fault arcs in power systems. *Eur. Trans. Electr. Power* **1991**, *1*, 55–60. [CrossRef]

32. Huang, N.E.; Shen, Z.; Long, S.R.; Wu, M.C.; Shih, H.H.; Zheng, Q.; Yen, N.C.; Tung, C.C.; Liu, H.H. The empirical mode decomposition and the Hilbert spectrum for nonlinear and non-stationary time series analysis. *Proc. R. Soc. Lond. Ser. A Math. Phys. Eng. Sci.* **1998**, *454*, 903–995. [CrossRef]

33. Ge, Y.; Sui, F.; Xiao, Y. Prediction methods for preventing single-phase reclosing on permanent fault. *IEEE Trans. Power Deliv.* **1989**, *4*, 114–121.

34. Jafarian, P.; Eskandari, H.; Sanaye-Pasand, M. Sizing Neutral Reactor Regardless of Line Length in Shunt Compensated Transmission Lines. In Proceedings of the 31th Power System Conference, Tehran, Iran, 24–26 October 2016.
35. Jafarian, P.; Eskandari, H.; Sanaye-Pasand, M. Application of universal neutral reactor in shunt compensated transmission lines: Feasibility study. *IET Gener. Transm. Distrib.* **2018**, *12*, 2181–2189. [CrossRef]
36. Zadeh, M.R.; Sanaye-Pasand, M.K. Investigation of neutral reactor performance in reducing secondary arc current. *IEEE Trans. Power Deliv.* **2008**, *23*, 2472–2479. [CrossRef]
37. Al-Shetwi, A.Q.; Hannan, M.A.; Jern, K.P.; Alkahtani, A.A.; PG Abas, A.E. Power Quality Assessment of Grid-Connected PV System in Compliance with the Recent Integration Requirements. *Electronics* **2020**, *9*, 366. [CrossRef]
38. Dias, O.; Magrin, F.; Tavares, M.C. Comparison of secondary arcs for reclosing applications. *IEEE Trans. Dielectr. Electr. Insul.* **2017**, *24*, 1592–1599. [CrossRef]

Publisher's Note: MDPI stays neutral with regard to jurisdictional claims in published maps and institutional affiliations.

Article

An Improved Inverse-Time Over-Current Protection Method for a Microgrid with Optimized Acceleration and Coordination

Liang Ji [1], Zhe Cao [2], Qiteng Hong [3], Xiao Chang [4], Yang Fu [1,*], Jiabing Shi [5], Yang Mi [1] and Zhenkun Li [1]

[1] School of Electrical Engineering, Shanghai University of Electric Power, Shanghai 200090, China; jiliang@shiep.edu.cn (L.J.); miyang@shiep.edu.cn (Y.M.); lizhenkun@shiep.edu.cn (Z.L.)
[2] State Grid Fengyang County Power Supply Company, Fengyang 233100, China; shdldxcaozhe@163.com
[3] Department of Electronic and Electrical Engineering, University of Strathclyde, Glasgow G1 1RD, UK; q.hong@strath.ac.uk
[4] State Grid Shanxi Electric Power Research Institute, Shanxi 030001, China; changxiao@sx.sgcc.com.cn
[5] State Grid Shanghai Qingpu Power Supply Company, Shanghai 200090, China; shdlsjb@163.com
* Correspondence: fuyang@shiep.edu.cn

Received: 30 August 2020; Accepted: 24 October 2020; Published: 2 November 2020

Abstract: This paper presents an improved inverse-time over-current protection method based on the compound fault acceleration factor and the beetle antennae search (BAS) optimization method for a microgrid. The proposed method can not only significantly increase the operation speed of the inverse-time over-current protection but also improve the protection coordination by considering the possible influential factors in terms of microgrid operation modes, distributed generation (DG) integration status, fault types, and positions, which are marked as the most challenging problems for over-current protection of a microgrid. In this paper, a new Time Dial Setting (TDS) of inverse-time protection is developed by applying a compound fault acceleration factor, which can notably accelerate the speed of protection by using low-voltage and short-circuit impedance during the fault. In order to improve the protection coordination, the BAS algorithm is then used to optimize the protection parameters of the pick-up current, TDS, and the inverse time curve shape coefficient. Finally, case studies and various evaluations based on DIgSILENT/Power Factory are carried out to illustrate the effectiveness of the proposed method.

Keywords: microgrid; distributed generation; inverse-time over-current protection; coordination optimization

---

1. Introduction

Nowadays, with the increasing demand for electric energy, distributed generations (DGs), such as photovoltaics, wind turbines, and fuel cells, play significant roles in power systems [1,2]. Since the DG in the microgrid is close to the load, the line loss is reduced and the energy efficiency is improved, which makes a great contribution to the reduction of carbon dioxide emissions in the world [3,4]. As the majority of DGs are connected to the grid via power electronic inverters, which is called inverter-interfaced DG (IIDG), new challenges have risen in energy management, planning, and design, as well as control and protection. Effective microgrid protection is the primary prerequisite for the reliable operation of a microgrid [5,6].

Considering the unique fault characteristics, the conventional protection methods used in distribution systems are considered no longer sufficiently effective [7]. Firstly, the fault current level is strict by the current limiter of the inverter, which introduces challenges to the majority of conventional current-based protection methods [8]. Secondly, a microgrid has two different operational modes, i.e., a grid-connected mode and islanding mode, which causes notably different fault levels [9]. Thirdly, as microgrids can have multiple DGs, the fault current direction may be changed, which brings difficulties in the relay setting and coordination [10]. Furthermore, the inverter control methods alter the relationship of the output voltage and current compared with the synchronous generator and inherently increases the risk of compromising the performance of the conventional protection method [11].

Many researchers have put forward efforts to improve the performance of the microgrid protection system. Some of them focus on protection with the communication channel [12,13]. Slemaisardoo et al. [12] proposed a differential protection method using a non-nominal frequency current during the fault, which can better detect the microgrid fault than conventional over-current protection. Aghdam et al. [13] proposed a differential protection method based on variable tripping times, and a multi-agent protection scheme was designed to improve the coordination of adjacent relays. Except for differential protection, some papers [14–16] used the data from multiple measurement points to classify and optimize the corresponding protection parameters based on the state change of the microgrid topology and DGs. Communication-based methods are reasonable solutions for microgrid protection. However, the reliability of this type of protection highly relies on the communication facilities and performance; also, it is not an economic solution.

Some research focused on protection methods that do not rely on communication. In [17], a new directional relay using amplitude and phase of sequence components in the network is proposed. In [18], a new zero-sequence direction protection for microgrid ground faults is proposed. Huang et al. [19] proposed an inverse-time impedance protection method, which is not affected by changes in the microgrid short-circuit level. In [20], an adaptive distance protection method based on the auxiliary coefficient is proposed to solve the influence of DG on the measured impedance of the relay. Some researchers proposed improved over-current protection methods by considering the fault characteristics of the microgrid [21–23]. Muda and Jena [21] proposed an adaptive over-current protection method by increasing the fault current during the fault. The superposition current of both the positive and negative sequences was applied to amplify the value of the fault current. In [22], an adaptive over-current protection method that diagnosed the microgrid operating mode from the voltage analysis is proposed. El-Naily et al. [23] designed a new pickup current constraint based on the influence of DG on over-current protection so as to overcome the low-fault-level problem. Besides, some papers proposed methods using fault current limiters (FCLs) to limit the short-circuit level of the microgrid [24,25]. In [24], FCLs are used to limit the fault current contribution of the main grid, and the genetic algorithm is used to solve the optimal coordination of the protection. In [25], a new hybrid method that combined the Cuckoo optimization algorithm and linear programming (COA–LP) was applied to solve the coordination of the microgrid protection. Although the use of FCLs can ensure the reliability and selectivity of traditional over-current protection, the investment of additional equipment increases the cost of protection and lacks economic efficiency.

With the development of artificial intelligence technology, some smart protection methods were also developed in recent years [26–28]. Kar et al. [26] proposed a differential protection scheme based on data mining to adapt to the operating mode and topology changes of the microgrid. In [27], an adaptive protection scheme based on machine learning was proposed, which can adaptively modify the protection parameters for different operating conditions. Mishra and Rout [28] proposes a differential protection scheme based on Hilbert–Huang transform (HHT) and machine learning, in which HHT was used for feature extraction and machine learning used to classify the fault. These methods can effectively solve the relay setting and coordination problems, but the reliability and fastness of some smart algorithms are difficult to guarantee.

Although massive smart and sophisticated protection methods have been developed, the simple over-current relay (OCR), which has the characteristics of good performance, simple principles, and a low cost, is still widely used in low-voltage power grids. In practice, a lot of existing microgrid projects utilize the over-current relay as their main and backup protection methods [29]. However, some challenges are still not overcome. Firstly, the over-current relays utilized in a microgrid are mainly time-inverse over-current relays, the operation time of which is inversely proportioned to the fault current. Due to the low fault-current level of the microgrid, the operational speed of this type of relay is difficult to be satisfactory. Secondly, a microgrid typically has multiple DGs located in different branches, which may bring difficulty in relay coordination. Furthermore, the alteration of the operation modes of the microgrid will notably change the fault level and challenge the relay settings.

Therefore, to overcome the limitations of the conventional over-current protection methods discussed above, this paper proposes an improved inverse-time over-current (I-ITOC) protection method. A compound fault acceleration factor based on low voltage and the measured impedance was developed to improve the speed of the relay. Then, the coordination of protection is optimized by using the beetle antenna search (BAS) algorithm. Compared with the conventional over-current method, the proposed method notably improves the operation speed. Furthermore, as the proposed method does not require extra devices, it is potentially more economic and easier to implement in the field.

The rest of the paper is organized as follows: In Section 2, the compound fault acceleration factor and the I-ITOC protection method are explained. The optimal configuration of the protection parameters is described in Section 3, and case studies with simulation comparison analyses are presented in Section 4.

2. Improved Inverse-Time Over-Current Protection Method Based on the Compound Fault Acceleration Factor

2.1. Introduction to the Inverse-Time Over-Current Relay

The inverse-time over-current relay (ITOCR) has the ability to reflect the severity of faults, and its operation time is inversely proportioned to the fault current. According to the standard IEC 60255 [30], the operation characteristic equation of the ITOCR is defined as follows:

$$ t = \frac{A}{\left(\frac{I_f}{I_p}\right)^\alpha - 1} \times \text{TDS} \tag{1} $$

where t is the relay's operation time, A is the constant coefficient, α is the inverse-time curve shape coefficient, I_f is the magnitude of the fault current measured by the relay, Time Dial Setting (TDS) is the time dial setting, and I_p is the setting of the relay's pickup current.

However, due to the changes in the operating mode of the microgrid, the limitations of the power electronics in contributing to the fault current in the microgrid, and the significant impact of the DGs' control strategy and capacity on the fault output current, there are significant differences in fault current values in the microgrid under different modes. Since the operation time of a conventional ITOCR is closely related to the fault current, the large variations in fault level will significantly affect the performance of the conventional ITOCR. Therefore, in order to ensure the satisfactory speed of the ITOCR in different modes of the microgrid, it is necessary to improve the conventional ITOC protection method.

2.2. Development of the Compound Fault Acceleration Factor for Over-Current Protection of a Microgrid

When the microgrid is operating in the grid-connected mode, the bus voltage and frequency are regulated by the upstream main grid. When the microgrid is switched to the islanded mode, one or some main DGs will be used to maintain the bus voltage and frequency stability. In the event of a fault, the closer the fault point is to the relay installation point, the more serious the voltage drop of the relay will be. Thus, the voltage drop of the relays can reflect the distance from the fault point to the installation of the relay.

As shown in Figure 1, when one fault occurs downstream of the relay R2 (point f), constructing the coefficient of the fault voltage based on the characteristics of the bus fault voltage, and the fault voltage coefficient U_i^* of the relay Ri, can be defined in Equation (2):

$$U_i^* = 1 - \left| \frac{U_i^{\text{fault}} - U_i^{\text{prefault}}}{U_i^{\text{prefault}}} \right| \tag{2}$$

where U_i^{prefault} and U_i^{fault} are the pre-fault voltage and the fault voltage to the relay Ri, respectively. The fault position on the line influences the value of U_i^*. The closer the relay is to the point of fault, the smaller U_i^* is, and U_i^* is less than 1 for all possible scenarios.

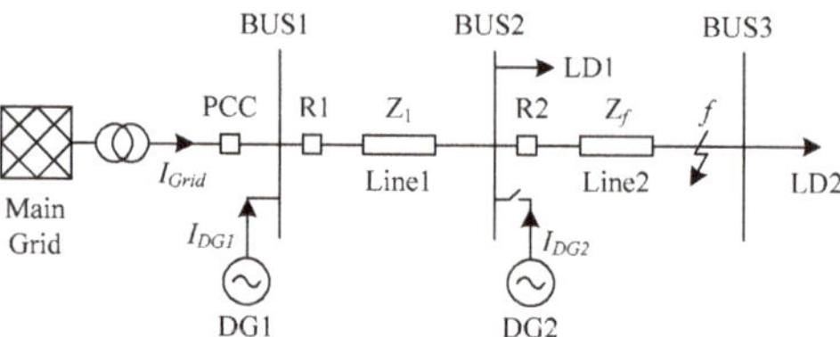

Figure 1. Simplified microgrid model.

In addition, according to the principle of distance protection, the measured impedance of the relay reflects the distance between the relay and the point of fault, and the fault impedance coefficient Z_i^* can be constructed as follows:

$$Z_i^* = \frac{|Z_L|}{|Z_{Ri}|} \tag{3}$$

where Z_L is the line impedance, which is the total impedance between the bus PCC (BUS1) to the end of the branch (BUS3). Z_{Ri} is the measured impedance of the relay Ri. Please note that calculation of Z_{Ri} changes with different fault types; also, the closer the relay is to the point of fault, the larger the fault impedance coefficient Z_i^* is. Z_i^* is greater than 1 for all possible scenarios.

By combing Equations (2) and (3), the operation characteristic equation and new TDS of the I-ITOC protection method is obtained in (4) and (5), where M_i in (6) represents the compound fault acceleration factor.

$$t_i = \frac{A}{\left(\dfrac{I_{f \cdot i}}{I_{p \cdot i}} \right)^{\alpha_i} - 1} \times \text{TDS}_i^* \tag{4}$$

$$\text{TDS}_i^* = \text{TDS}_i \times M_i \tag{5}$$

$$M_i = \frac{U_i^*}{Z_i^*} \tag{6}$$

Compared with the conventional ITOCR, the speed of the improved inverse-time over-current relay (ITOCR) can be improved as M_i is less than 1 for the majority of scenarios. The following section, Section 2.3, will discuss a few scenarios in which M_i is greater than 1. In the case of the islanded mode of the microgrid, due to the influence of the DG capacity, the DG's maintain bus voltage capability is weaker than that of the main grid. Therefore, the compound fault acceleration factor M_i in the islanded mode is smaller than that of the grid-connected mode. Inherently, the influence of a low fault level on the protection operation time during the island mode can be further reduced.

2.3. Effect of DG on the Compound Fault Acceleration Factor

The presence of DGs in the microgrid can change the fault current flowing through the relay, which will result in a different operation time by applying the ITOC. Besides, the fault current of the

DG may also affect the fault measurement impedance of the backup relay. In Figure 1, when a fault occurs downstream of the relay R2, the impedance measured by the relays R1 and R2 can be obtained by Equations (7) and (8):

$$Z_{R1} = Z_1 + Z_f + Z_{DG}^* \tag{7}$$

$$Z_{R2} = Z_f \tag{8}$$

where

$$Z_{DG}^* = \frac{I_{DG2}}{I_{BUS1}} \times Z_f \tag{9}$$

In Equations (7) and (8), Z_{R1} and Z_{R2} are the measured impedance of the relay R1 and R2, Z_f is the impedance between relay R2 and fault point f, and Z_{DG}^* is the contribution impedance of DG2's fault current to relay R1. In Equation (9), I_{BUS1} is the fault current flowing through BUS1 and I_{DG2} is the fault current of the DG2.

When the fault occurs in the grid-connected mode, I_{BUS1} will typically be much greater than I_{DG2}, and Z_{DG}^* can be ignored. When the fault occurs in the microgrid islanded mode, the magnitude of $|I_{DG2}/I_{BUS1}|$ is affected by the DG control strategy and the severity of the fault. Since the phase angle of $|I_{DG2}/I_{BUS1}|$ is between $-90°$ and $90°$ [19], Z_{R1} is still greater than Z_{R2}, and the fault impedance coefficient Z_1^* and Z_2^* measured by the relays R1 and R2 still satisfies the criteria that Z_2^* is greater than Z_1^*.

However, when the capacity of the DG is relatively large, the fault current of the DG may cause may cause the protection measurement impedance Z_{Ri} to be greater than the set impedance Z_L, inherently affecting the speed of the protection. Therefore, the compound fault acceleration factor M_i and the I-ITOC protection method is improved, which can be expressed in Equations (10) and (11).

$$M_i = \frac{U_i^*}{1 + Z_i^*} \tag{10}$$

where

$$Z_i^* = \begin{cases} 1 & Z_i^* < 1 \\ |Z_L/Z_{Ri}| & Z_i^* \geq 1 \end{cases} \tag{11}$$

In order to maintain the acceleration effect of M_i when the contribution of DG to the protection measurement impedance is too large, the denominator of M_i is changed, and 1 is taken when the fault impedance coefficient Z_i^* is less than 1.

3. Protection Coordination Optimization Based on the Beetle Antennae Search Algorithm

Due to the unique fault characteristics in the microgrids and the influence of the DG, the conventional settings of the ITOCR cannot be directly applied for effective microgrid protection. Figure 2a,b are the protection operation characteristics curve of the conventional ITOCR under different operation modes of the microgrid.

As shown in Figure 2a, if the settings are configured based on the grid-connected operational mode, the speed of the relay in the islanded mode could be too slow due to the reduced fault level; as shown in Figure 2b, if the settings are calculated based on the islanded operational mode, the coordination time interval (*CTI*) of the adjacent relays in the grid-connected mode could be too small to ensure the coordination of protection. In addition, the contribution of the DG to the fault current flowing through the relay also affects the coordination of the relay. Therefore, in order to ensure the selectivity and speed of I-ITOCR in different operation modes of the microgrid, it is necessary to optimize the parameters of each relay.

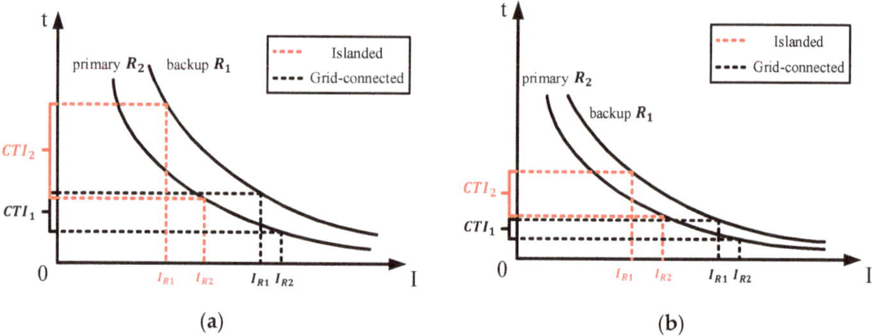

Figure 2. Operation time in different modes of the microgrid: (**a**) Parameters are set according to the grid-connected mode; (**b**) parameters are set according to the islanded mode.

3.1. Coordinated Optimization of the I-ITOCRs

The I-ITOCR operating characteristic equation is shown in Equation (6). In general, parameters A and α are fixed for a certain relay and the parameters $I_{p\cdot i}$ and TDS_i need to be set. In this paper, in order to reduce the impact of current changes and achieve optimal coordination, the index α_i is also added as the optimization variable. The objective function of the optimization problem is formulated in Equation (12), which represents the sum of the operation time for I-IOCR considering both the grid-connected and islanded operation modes of the microgrid.

$$\underset{I_p,\alpha,TDS}{\text{Min}}\ F = \sum_{m=1}^{2}\sum_{n=1}^{4}\left(\sum_{k=1}^{K}\left(t_{pr\cdot k}^{i} + \sum_{j=1}^{2}t_{bc\cdot k}^{ij}\right)\right) \tag{12}$$

where m denotes the operation mode of the microgrid, and the numbers 1 and 2, respectively, indicate the grid-connected and islanded modes of the microgrid; n represents different fault types; k represents the different fault conditions, here being the first and end fault of each line; i represents the primary relay and j represents the backup relay to the primary relay I; and t denotes the operation time of the relays, with $t_{pr\cdot k}^{i}$ denoting the operation time of the primary relay and $t_{bc\cdot k}^{ij}$ denoting the operation time of the backup relay.

The total operation time of the I-IOCRs is reduced by obeying their constraints as follows.

3.1.1. Protection Coordination Constraints

When a fault occurs in the network, in order to ensure the selectivity of the relay, there should be a reasonable *CTI* between the backup and primary relay operation time and this relationship also exists between the remaining protection and backup protection operation time. The *CTI* can be expressed as Equations (13) and (14):

$$CTI_{i\cdot 1} = t_{bc\cdot k}^{i1} - t_{pr\cdot k}^{i} \tag{13}$$

$$CTI_{i\cdot 2} = t_{bc\cdot k}^{i2} - t_{bc\cdot k}^{i1} \tag{14}$$

The *CTI* should have a minimum value by considering the operation time of the circuit breaker and a safety margin, where an excessive *CTI* is unfavorable to the stability of the system [31]. Therefore, the $CTI_{i\cdot 1}$ is limited to between 0.2 and 0.5, and the $CTI_{i\cdot 2}$ is greater than 0.2 [32]. It can be represented as follows:

$$CTI_{\min} \leq CTI_{i\cdot 1} \leq CTI_{\max} \tag{15}$$

$$CTI_{\min} \leq CTI_{i\cdot 2} \tag{16}$$

3.1.2. Protection Operation Time Constraint

Considering the speed of the relay and the stability of the system, the maximum and minimum operation time of the relay must be limited. For all the I-ITOCRs, the protection operation time is set between 0 and 2 s. The limit of the protection operation time is as shown in Equations (17) and (18):

$$t_{min} \leq t^{i}_{pr \cdot k} \leq t_{max} \tag{17}$$

$$t_{min} \leq t^{ij}_{bc \cdot k} \leq t_{max} \tag{18}$$

3.1.3. Protection Parameter Constraint

In this part, the constraint of the pickup current $I_{p \cdot i}$, the Time Dial Setting TDS_i, and the index α_i are introduced.

The range of values for each relay's pickup current is determined by the network. According to the requirements of the IEC standard, the pickup current must be greater than the maximum load current $I_{L \cdot max}$ of the line, which is normally set as $I_{p \cdot i} > 1.1 I_{L \cdot max}$ [23]. In order to ensure the reliability of the protection, the pickup current must be smaller than the minimum fault current $I_{f \cdot min}$, which is the fault current of the two-phase fault at the remote end of the feeder. The TDS_i is generally set between 0.05 and 11 s [33]. According to Standard IEC 60255, the inverse time type α_i is usually between 0.02 and 13.5 [34]. Finally, the lower and upper bounds of the I-ITOCR's parameters are given in (19)–(21).

$$\alpha_{min} < \alpha_i < \alpha_{max} \tag{19}$$

$$I_p^{min} \leq I_{p \cdot i} \leq I_p^{max} \tag{20}$$

$$TDS_{min} \leq TDS_i \leq TDS_{max} \tag{21}$$

3.2. Parameter Optimization of the I-ITOCR Based on the Beetle Antenna Search Algorithm

According to the microgrid network structure and the fault condition, the objective function (12) and the constraints (13)–(16) can be obtained. Next, considering the constraints (17)–(21) of the parameter variables, the optimal parameter configuration for each relay can be determined. Since the above problem is a nonlinear optimization problem, the solving process is complicated and difficult. Therefore, in this section, the BAS algorithm is used to solve this nonlinear optimization problem to obtain the optimal configuration of the I-ITOCR's parameters.

The BAS algorithm is a meta-heuristic algorithm for multi-objective function optimization, which imitates the perception function of beetle antennae to judge the fitness of local areas around itself and guides individuals to move to the global optimal solution through the optimal solution of local areas [35]. Compared with the particle swarm optimization algorithm, the BAS algorithm only needs one individual, which greatly reduces the computational complexity.

The algorithm flow of the BAS is shown in Figure 3, and its main steps are explained as follows:

Step 1: Randomly generate the direction vector $\vec{b}$ of the beetle antennae and normalize it.

$$\vec{b} = \frac{rands(g, 1)}{\|rands(g, 1)\|} \tag{22}$$

where *rands* denotes a random function and g presents the dimensions of the position. Here, the size of g is related to the number of relays to be optimized.

Step 2: Calculate the left and right spatial coordinates X_l and X_r of the antennae according to the initial search distance d^t and $\vec{b}$.

$$X_l = X^t - d^t \vec{b} \tag{23}$$

$$X_r = X^t + d^t \vec{b} \tag{24}$$

where X^t is the position of the beetle at the tth iteration, and X^t can be expressed as Equation (25):

$$X^t = \left[I_{p\cdot1}^t, \alpha_1^t, \text{TDS}_1^t, \cdots, I_{p\cdot w}^t, \alpha_w^t, \text{TDS}_w^t \right]^T \tag{25}$$

where w is the number of relays to optimize.

Step 3: Calculate the odor intensity $f(X_l)$ and $f(X_r)$ based on the spatial coordinates and the fitness function $f(X)$.

$$f(X_l) = F(X_l) \tag{26}$$

$$f(X_r) = F(X_r) \tag{27}$$

where $F(.)$ is the objective function shown in Equation (12).

Step 4: Determine the position x^{t+1} of the next moment of the beetle according to the fitness function value.

$$X^{t+1} = X^t - \delta^t \vec{b}\, sign(f(X_l) - f(X_r)) \tag{28}$$

where δ^t represents the step size and sign(.) represents a sign function.

Step 5: Update the step size δ^t and search distance d^t.

Step 6: Iterate to the maximum number of iterations and output the result.

Through the above steps, the optimal parameters $I_{p\cdot i}$, TDS_i, and α_i of each relay in the microgrid can be obtained, and the coordination problem of the microgrid protection can be solved.

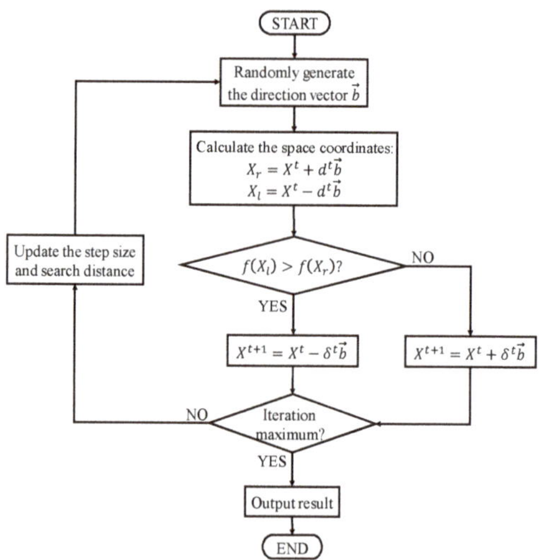

Figure 3. Flowchart of the beetle antenna search (BAS) algorithm.

3.3. Classification of Microgrid Scenarios Based on DG Status

Since the connection status of the DG in the microgrid affects the fault current in the network, it is necessary to classify the scenarios of the microgrid according to whether the DG is connected and obtain the optimal protection parameters under the corresponding scenarios. Please note that the information about the connection status of DG can be obtained from the central controller of the microgrid in practice [23].

Figure 4 shows a typical microgrid structure which contains 4 DGs and two branches. The classification of the microgrid scenarios follows the following criteria. First of all, it is not necessary to consider the PCC connection status, because the parameter optimization can comprehensively consider the microgrid operation modes, i.e., the grid-connected and islanded mode. Secondly, during the microgrid grid-connect operation mode, DGs apply conventional PQ control. During the microgrid island operation mode, the widely used master–slave control strategy is applied. DG1, which is assumed as a stable resource, is chosen as the unit to maintain the microgrid voltage and frequency. Finally, the scenarios of the microgrid are classified according to the state changes of the remaining DGs.

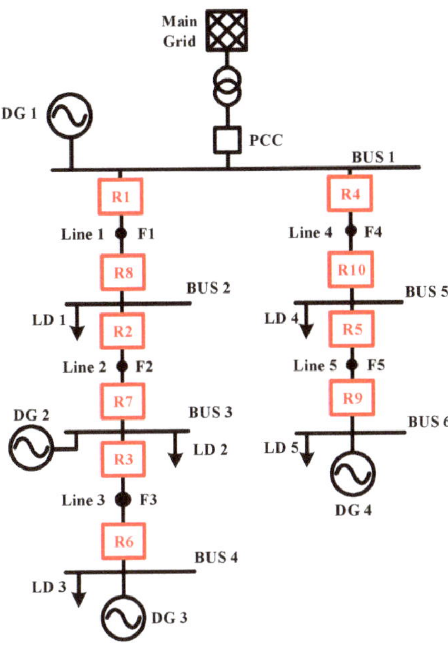

Figure 4. Schematic diagram of the microgrid.

According to the above classification principle, the microgrid network shown in Figure 4 can be divided into eight scenarios, as shown in Table 1. Relays R1–R10 are directional over-current relays. In different modes of the microgrid, the fault currents flowing through Relays R1, R2, R3, R4, and R5 are obviously different. It is necessary to optimize the parameters according to the above optimization process. The fault current flowing through Relays R6, R7, R8, R9, and R10 is less affected by the change in the operating mode of the microgrid, and the parameters can be configured according to the conventional protection method, which is beyond the scope of this paper.

Table 1. Scenarios classification of the microgrid.

Scenarios	DG2	DG3	DG4
1	OFF	OFF	OFF
2	ON	ON	OFF
3	ON	OFF	ON
4	ON	OFF	OFF
5	OFF	OFF	ON
6	OFF	ON	OFF
7	OFF	ON	ON
8	ON	ON	ON

4. Simulation Results

The proposed method based on the compound acceleration factor and BAS optimization algorithm aim to improve the speed and coordination of the microgrid protection. The impacts of negative factors on the conventional ITOCR in terms of a low fault level, microgrid operation mode, DG positions, and DG integration status have been notably reduced. In this section, the speed and coordination of the proposed protection method are evaluated. The possible negative influencing factors on the performance of the proposed method, such as fault types, fault positions, and microgrid operation modes, are evaluated. Besides, the comparison between the proposed method and the conventional method is also included in this section.

In the simulation, the model of a typical microgrid shown in Figure 4 is established in DIgSILENT/Power Factory. The detailed parameters of the microgrid are shown in Table 2. All DGs adopted the PQ control during the grid-connected operation mode of the microgrid. During the islanded operation mode, the microgrid applies the master–slave control mode. DG1, which is assumed as the most stable and reliable resource in the microgrid, is denoted as the master DG and adopts the V/f control. In turn, the other DGs kept using the PQ control. The initial state of the DGs in the microgrid is Scenario 1 in Table 1. The change in DG status will be evaluated in this section. Due to the thermal limit of the power inverter, the fault current limit of the DGs is set as 2 p.u. For conventional ITOCRs, the parameters of A and α are set as 0.14 and 0.02, and the minimum allowed CTI is set to 0.3 s [22,32]. F1 to F5 in Figure 4 denotes the fault positions.

Table 2. Simulation parameters of the microgrid model.

Element	Parameter
DG1/DG2/DG3/DG4	1500/100/200/500 kW
Line Parameters and length	0.01 + j0.015 Ω/km, 1 km
Load1/Load2/Load3/Load4/Load5	300/100/200/200/400 kW
PCC Voltage/Frequency	0.4 kV/50 Hz

4.1. Optimized Configuration of the Protection Parameters

The settings parameters of the microgrid protection of Scenario 1 are optimized based on the methods described in Section 3. Table 3 shows the optimal protection parameters obtained according to the BAS algorithm optimization process shown in Figure 3.

Table 3. Optimal configuration of the protection parameters in Scenario 1.

Relay	$I_{p \cdot i}/I_{L \cdot max_i}$	TDS_i	α_i
R1	1.1277	0.5061	0.0211
R2	1.1308	0.7231	0.0201
R3	1.5788	0.7511	0.0214
R4	1.1105	8.7153	0.2981
R5	2.9282	0.1495	0.0262

4.2. Evaluation of the Acceleration Capability of the Proposed Method

In this section, the acceleration capability of the proposed method is evaluated with different microgrid operating modes, fault types, and positions, and the comparison with conventional protection methods is carried out.

4.2.1. Evaluations in the Grid-Connected Mode of the Microgrid

(a) Different fault positions: In this case, the simulations are performed for different fault positions in the grid-connected mode. The three-phase faults are applied in this case, while the evaluations of different fault types will be discussed in the following section.

When fault F3 of Line 3 occurred, the fault current is measured as 5.2553 kA, and the acceleration factor M_i is calculated to be 0.0417 according to (10). According to the protection parameters in Table 3, the action delay of R3 is 0.0499 s. Figure 5 shows the voltage change of the bus when occurring at F3 of Line 3. From Figure 5, it is found that the operation time of the conventional primary relay is 0.1096 s, and the improved primary relay is 0.0597 s faster than the traditional relay. Table 4 shows the simulation results with different fault positions. The simulation results prove that the speed of the improved protection method can be guaranteed with different fault positions, and both the primary and backup relay's operation times of the I-ITOCR are faster than those of the ITOCR.

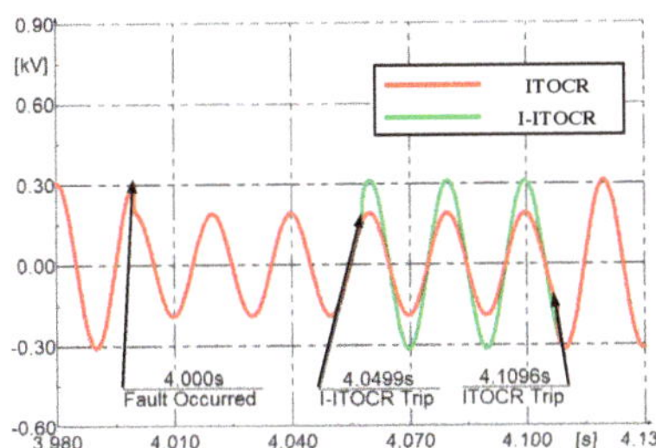

Figure 5. Bus voltage during the fault at F3.

Table 4. Operation times of the different fault positions in the grid-connected mode.

Fault Position	Relay Operational Time (s)			
	I-ITOCR		ITOCR	
	PR	BR	PR	BR
F1	0.1110	—	0.4678	—
F2	0.0701	0.4259	0.3549	0.7307
F3	0.0499	0.3799	0.1096	0.4437
F4	0.0837	—	0.3094	—
F5	0.0204	0.3789	0.1179	0.4828

Note: PR and BR represent primary relay and backup relay, respectively.

(b) Different fault types: This section evaluates the speed of the proposed protection method with different fault types in the grid-connected mode. Table 5 shows the operation times of the I-ITOCRs and ITOCRs. The fault is set at the F2 of Line 2. In the table, A, B, C, and D represent a single-phase-to-ground fault, a two-phase fault, a two-phase-to-ground fault, and three-phase fault, respectively. The operation time for the improved primary relays are all less than 0.15 s, and the conventional primary relays are all greater than 0.35 s, which proves that the I-ITOCR can adopt all the fault types and is faster than the ITOCRs under different fault types.

Table 5. Operation times of the different fault types in the grid-connected mode.

Fault Type	Relay Operational Time (s)			
	I-ITOCR		ITOCR	
	PR	BR	PR	BR
A	0.1409	0.5264	0.3556	0.7297
B	0.0676	0.4151	0.3729	0.7705
C	0.0640	0.3947	0.3552	0.7292
D	0.0701	0.4259	0.3549	0.7307

4.2.2. Evaluations in the Islanded Mode of the Microgrid

(a) Different fault positions: This case tests the speed of the proposed protection method for different fault position in the islanded mode. Please note that the three-phase fault is included in this case; other fault types will be discussed in the following section.

When fault F3 of Line 3 occurred, the fault current is 3.1970 kA, and the acceleration factor M_i is 0.0412 according to (10). According to the protection parameters in Table 3, the action delay of R3 is 0.0568 s. Figure 6 shows the voltage of the I-ITOCR and ITOCR of BUS2 when a fault occurs at F3 of Line 3. As can be seen from Figure 6, the operation time of the conventional primary relay is 0.1650 s, and the action delay of the improved primary relay is much shorter than that of the conventional protection method.

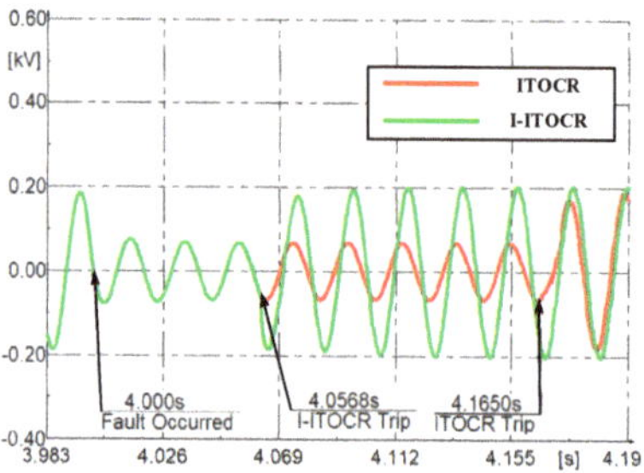

Figure 6. Bus voltage during the fault at F3.

Table 6 shows the simulation results of the other faults in this case. Note that, since the fault level of the microgrid is reduced, when the fault occurs at the F1 point, the operation time of the conventional primary relay is 1.1519 s, and the operation time for the improved primary relay is only 0.0728 s, which is much smaller than the conventional protection method. In addition, the primary and backup relay's operation time at other points of the I-ITOCR is also notably faster than the ITOCR. Therefore, the improved method has a better speed in the islanded mode of the microgrid.

Table 6. Operation times of the different fault positions in the islanded mode.

Fault Position	Relay Operational Time (s)			
	I-ITOCR		ITOCR	
	PR	BR	PR	BR
F1	0.0728	—	1.1519	—
F2	0.0667	0.5033	0.6013	1.5919
F3	0.0568	0.4419	0.1650	0.7408
F4	0.0715	—	0.7591	—
F5	0.0389	0.5181	0.2155	1.0478

(b) Different fault types: In this case, the speed of the improved protection method for different types of faults in the islanded mode is tested. Table 7 shows the operation times of the ITOCRs and I-ITOCRs when different types of faults occur at the F3 of Line 3 of the microgrid. As can be seen from Table 7, the operation time for the improved primary relays are all less than 0.15 s, and the conventional primary relays are all greater than 0.16 s, which proves that the I-ITOCR is faster than the ITOCRs under different failure types in the islanded mode of the microgrid.

Table 7. Operation times of the different fault types in the islanded mode.

Fault Ttype	Relay Operational Time (s)			
	I-ITOCR		ITOCR	
	PR	BR	PR	BR
A	0.1490	0.5409	0.1677	0.7508
B	0.0481	0.4161	0.1915	0.9061
C	0.0430	0.3379	0.1667	0.6681
D	0.0568	0.4419	0.1650	0.7408

4.3. Evaluations of Protection Coordination of the Proposed Method

In this section, the coordination of the protection methods is evaluated via the aspects of different operating modes, fault positions and types, and DG integration status. In this simulation, the traditional protection parameters are set according to the islanded mode of the microgrid.

4.3.1. Evaluations in the Grid-Connected Mode of the Microgrid

(a) Different fault positions: The simulations are performed with different fault position in the grid-connected mode. The fault type in this section is set as a three-phase fault; other fault types are discussed in the following section.

Table 8 shows the simulation results of the different fault positions. From the table, it is found that the *CTIs* between the primary and backup relays of the conventional protection are all less than 0.195 s, which causes the backup relay to remain active after the primary relay trips (*CTI* < 0.2). As a result, the selectivity of the protection cannot be guaranteed.

Table 8. Operation times of the different fault positions in the grid-connected mode

Fault Position	Relay Operational Time (s)					
	I-ITOCR			ITOCR		
	PR	BR	CTI	PR	BR	CTI
F1	0.1110	—	—	0.2162	—	—
F2	0.0701	0.4259	0.3558	0.2126	0.3377	0.1251
F3	0.0499	0.3799	0.3300	0.0724	0.2658	0.1934
F4	0.0837	—	—	0.1394	—	—
F5	0.0204	0.3789	0.3585	0.0621	0.2175	0.1554

For the proposed I-ITOCR method, the *CTIs* between the primary and backup relays of the improved protection method is all around 0.35 s, which fully satisfy the coordination requirement (0.2 < *CTI* < 0.5). Therefore, improved protection can ensure the selectivity for different fault positions in the grid-connected mode.

(b) Different fault types: This section evaluates the coordination of the proposed protection method for different types of faults in the grid-connected mode. Table 9 shows the operation times of the ITOCRs and I-ITOCRs with different fault types occurring at the F2 of Line 2. From the results, it is found that the *CTIs* of the conventional ITOCR are all less than 0.14 s, which do not satisfy the coordination requirement (*CTI* < 0.2). In turn, the *CTI* of the improved I-ITOCR is all in the range (0.2 < *CTI* < 0.5) for different fault types, which promises the protection coordination and selectivity.

Table 9. Operation times of the different fault types in the grid-connected mode.

Fault Type	Relay Operational Time (s)					
	I-ITOCR			ITOCR		
	PR	BR	*CTI*	PR	BR	*CTI*
A	0.1409	0.5264	0.3855	0.2131	0.3372	0.1241
B	0.0676	0.4151	0.3475	0.2234	0.3561	0.1327
C	0.0640	0.3947	0.3307	0.2128	0.3382	0.1254
D	0.0701	0.4259	0.3558	0.2126	0.3377	0.1251

4.3.2. Evaluations in the Islanded Mode of the Microgrid

(a) Different fault positions: In this case, the simulation is performed for different fault position in the islanded mode. Table 10 shows the simulation results of the example, and the faults are all three-phase faults; other fault types are discussed in the following section. As can be seen from the simulation results, the *CTI* between the primary and backup relays of the improved protection and conventional protection methods meet the condition (0.2 < *CTI* < 0.5), and the selectivity of these protections is not lost for different fault position in the islanded mode.

Table 10. Operation times of the different fault positions in the islanded mode.

Fault Position	Relay Operational Time (s)					
	I-ITOCR			ITOCR		
	PR	BR	*CTI*	PR	BR	*CTI*
F1	0.0728	—	—	0.5322	—	—
F2	0.0667	0.5033	0.4366	0.3602	0.7355	0.3753
F3	0.0568	0.4419	0.3851	0.1088	0.4437	0.3349
F4	0.0715	—	—	0.3419	—	—
F5	0.0389	0.5181	0.4792	0.1135	0.4718	0.3583

(b) Different fault types: This section evaluates the coordination of the proposed protection method for different types of faults in the islanded mode. Table 11 shows the operation time of ITOCRs and I-ITOCRs with different fault types occurring at the F2 of Line 2. From the results, it is found that the CTIs of the I-ITOCR and ITOCR are all in the range (0.2<CTI<0.5) for different fault types, which promises the protection coordination and selectivity.

Table 11. Operation times of the different fault types in the grid-connected mode.

Fault Type	Relay Operational Time (s)					
	I-ITOCR			ITOCR		
	PR	BR	*CTI*	PR	BR	*CTI*
A	0.1578	0.5055	0.3477	0.3633	0.7383	0.3750
B	0.0525	0.5091	0.4566	0.4464	0.8515	0.4051
C	0.0475	0.4450	0.3975	0.3220	0.6140	0.2920
D	0.0667	0.5033	0.4366	0.3602	0.7355	0.3753

4.3.3. Evaluations of the DG Connection States

In this case, the simulations were carried out with different DG states in the microgrid and the protection parameter settings in the different scenarios were consistent with the method of Section A. Table 12 shows the operation time of the I-ITOCRs when the fault occurs at F3 of Line 3 of the microgrid for eight scenarios; the faults are all three-phase faults, and G and I represent the grid-connected mode and islanded mode of the microgrid, respectively. As can be seen from the simulation results, the improved protection method is not affected by the change of DG state, and the selectivity of the relay can be guaranteed under different scenarios and different operating modes of the microgrid ($0.2 < CTI < 0.5$).

Table 12. Operating times of the I-ITOCRs in the different scenarios and different operating modes of the microgrid.

Fault Position		Relay Operational Time (s)		
		PR	BR	*CTI*
Scenario 2	G	0.0027	0.3772	0.3745
	I	0.0029	0.4491	0.4462
Scenario 3	G	0.0117	0.3812	0.3695
	I	0.0118	0.4111	0.3993
Scenario 4	G	0.0677	0.3942	0.3265
	I	0.0821	0.4617	0.3796
Scenario 5	G	0.0246	0.3969	0.3723
	I	0.0235	0.3973	0.3738
Scenario 6	G	0.0293	0.4304	0.4011
	I	0.0263	0.3886	0.3623
Scenario 7	G	0.0093	0.3853	0.3760
	I	0.0081	0.3420	0.3339
Scenario 8	G	0.0241	0.4713	0.4472
	I	0.0229	0.4590	0.4361

5. Conclusions

In this paper, an improved inverse-time over-current protection method based on the compound fault acceleration factor and BAS optimization algorithm is proposed. The new method notably increases the speed of the over-current protection and improves the coordination between the adjacent relays. The potential influencing factors on ITOCR protection, including the microgrid operation modes, fault position, and fault types, were evaluated. The simulation results show that the improved inverse-time over-current protection method can quickly and coordinately remove faults under different operation modes, different fault locations, and different fault types of the microgrid, which verifies the speed and coordination of the protection method, and is better than the traditional protection methods. Furthermore, as the proposed method does not require extra devices, it is potentially more economic and easier to implement in the field, thus offering a promising solution for effective microgrid protection.

Author Contributions: Author Contributions: Conceptualization, L.J. and Z.C.; methodology, L.J. and Z.C.; software, J.S. and Z.C.; validation, L.J., Z.C. and J.S.; formal analysis, Q.H.; investigation, X.C.; resources, Y.F.; data curation, Y.M.; writing—original draft preparation, L.J. and Z.C.; writing—review and editing, L.J., Y.F. and Z.C.; visualization, Z.L.; supervision, L.J. and Z.L.; project administration, Y.F. and Y.M.; funding acquisition, Y.F. All authors have read and agreed to the published version of the manuscript.

Funding: This research received no external funding.

Nomenclature

BAS	Beetle antennae search
DG	Distributed generation
TDS	Time Dial Setting
IIDG	Inverter-interfaced distributed generation
FCLs	Fault current limiters
HHT	Hilbert–Huang transform
OCR	Over-current relay
ITOC	Inverse-time over-current
ITOCR	Inverse-time over-current relay
CTI	Coordination time interval

References

1. Liu, B.; Zhuo, F.; Zhu, Y. System operation and energy management of a renewable energy-based DC micro-grid for high penetration depth application. *IEEE Trans. Smart Grid* **2015**, *6*, 1147–1155. [CrossRef]
2. Badal, F.R.; Das, P.; Sarker, S.K. A survey on control issues in renewable energy integration and microgrid. *Prot. Control Mod. Power Syst.* **2019**, *4*, 87–113. [CrossRef]
3. Teimourzadeh, S.; Aminifar, F.; Davarpanah, M. Macroprotections for microgrids: Toward a new protection paradigm subsequent to distributed energy resource integration. *IEEE Ind. Electron. Mag.* **2016**, *10*, 6–18. [CrossRef]
4. Zhang, F.; Mu, L. New protection scheme for internal fault of multi-microgrid. *Prot. Control Mod. Power Syst.* **2019**, *4*, 159–170. [CrossRef]
5. Chen, Z.; Pei, X.; Yang, M. A novel protection scheme for inverter-interfaced microgrid (IIM) operated in islanded mode. *IEEE Trans. Power Electron.* **2018**, *33*, 7684–7697. [CrossRef]
6. Kumar, D.S.; Savier, J.S.; Biju, S.S. Micro-synchrophasor based special protection scheme for distribution system automation in a smart city. *Prot. Control Mod. Power Syst.* **2020**, *5*, 97–110. [CrossRef]
7. Hooshyar, A.; Iravani, R. Microgrid protection. *Proc. IEEE* **2017**, *105*, 1332–1353. [CrossRef]
8. Beheshtaein, S.; Cuzner, R.; Savaghebi, M. Review on microgrids protection. *IET Gener. Transm. Distrib.* **2019**, *13*, 743–759. [CrossRef]
9. Haron, A.R.; Mohamed, A.; Shareef, H. A review on protection schemes and coordination techniques in microgrid system. *J. Appl. Sci.* **2012**, *12*, 101–112. [CrossRef]
10. Fani, B.; Bisheh, H.; Sadeghkhani, I. Protection coordination scheme for distribution networks with high penetration of photovoltaic generators. *IET Gener. Transm. Distrib.* **2018**, *12*, 1802–1814. [CrossRef]
11. Belwin, J.; Raja, R. A review on issues and approaches for microgrid protection. *J. Renew. Sustain. Energy Rev.* **2017**, *67*, 988–997.
12. Soleimanisardoo, A.; Karegar, H.K.; Zeineldin, H.H. Differential frequency protection scheme based on off-nominal frequency injections for inverter-based islanded microgrids. *IEEE Trans. Smart Grid* **2019**, *10*, 2107–2114. [CrossRef]
13. Aghdam, T.S.; Karegar, H.K.; Zeineldin, H.H. Variable tripping time differential protection for microgrids considering DG stability. *IEEE Trans. Smart Grid* **2019**, *10*, 2407–2415. [CrossRef]
14. Sharma, A.; Panigrahi, B.K. Phase fault protection scheme for reliable operation of microgrids. *IEEE Trans. Ind. Appl.* **2018**, *54*, 2646–2655. [CrossRef]
15. Zanjani, M.G.; Mazlumi, K.; Kamwa, I. Application of μPMUs for adaptive protection of over-current relays in microgrids. *IET Gener. Transm. Distrib.* **2018**, *12*, 4061–4068. [CrossRef]
16. Jain, R.; Lubkeman, D.L.; Lukic, S.M. Dynamic adaptive protection for distribution systems in grid-connected and islanded modes. *IEEE Trans. Power Deliv.* **2019**, *34*, 281–289. [CrossRef]
17. Hooshyar, A.; Iravani, R. A new directional element for microgrid protection. *IEEE Trans. Smart Grid* **2018**, *9*, 6862–6876. [CrossRef]
18. Mahamedi, B.; Zhu, J.G.; Eskandari, M. Protection of inverter-based microgrids from ground faults by an innovative directional element. *IET Gener. Transm. Distrib.* **2018**, *12*, 5918–5927. [CrossRef]
19. Huang, W.; Nengling, T.; Zheng, X. An impedance protection scheme for feeders of active distribution networks. *IEEE Trans. Power Deliv.* **2014**, *29*, 1591–1602. [CrossRef]

20. Jin, L.; Jiang, M.; Yang, G. Fault Analysis of Microgrid and Adaptive Distance Protection Based on Complex Wavelet Transform. In Proceedings of the International Power Electronics and Application Conference and Exposition, Shanghai, China, 5–8 November 2014; pp. 1–5.

21. Muda, H.; Jena, P. Superimposed adaptive sequence current based microgrid protection: A new technique. *IEEE Trans. Power Deliv.* **2017**, *32*, 757–767. [CrossRef]

22. Mahat, P.; Chen, Z.; Bakjensen, B. A simple adaptive 0ver-current protection of distribution systems with distributed generation. *IEEE Trans. Smart Grid* **2011**, *2*, 428–437. [CrossRef]

23. Elnaily, N.; Saad, S.M.; Hussein, T. A novel constraint and non-standard characteristics for optimal over-current relays coordination to enhance microgrid protection scheme. *IET Gener. Transm. Distrib.* **2019**, *13*, 780–793. [CrossRef]

24. Najy, W.K.A.; Zeineldin, H.H.; Woon, W.L. Optimal protection coordination for microgrids with grid-connected and islanded capability. *IEEE Trans. Ind. Electron.* **2013**, *60*, 1668–1677. [CrossRef]

25. Dehghanpour, E.; Karegar, H.K.; Kheirollahi, R. Optimal coordination of directional over-current relays in microgrids by using cuckoo-linear optimization algorithm and fault current limiter. *IEEE Trans. Smart Grid* **2018**, *9*, 1365–1375. [CrossRef]

26. Kar, S.; Samantaray, S.R.; Zadeh, M.D. Data-mining model based intelligent differential microgrid protection scheme. *IEEE Syst. J.* **2017**, *11*, 1161–1169. [CrossRef]

27. Lin, H.; Sun, K.; Tan, Z.; Liu, C.; Guerrero, J.M.; Vasquez, J.C. Adaptive protection combined with machine learning for microgrids. *IET Gener. Transm. Distrib.* **2019**, *13*, 770–779. [CrossRef]

28. Mishra, M.; Rout, P.K. Detection and classification of micro-grid faults based on HHT and machine learning techniques. *IET Gener. Transm. Distrib.* **2018**, *12*, 388–397. [CrossRef]

29. Shiles, J.; Wong, E.; Rao, S. Microgrid protection: An overview of protection strategies in North American microgrid projects. In Proceedings of the IEEE Power and Energy Society General Meeting, Chicago, IL, USA, 16–20 July 2017; pp. 1–5.

30. IEC. *Measuring Relays and Protection Equipment—Part 151: Functional Requirements for Over/Under Current Protection, IEC 60255-151 Edition 1.0*; IEC: Geneva, Switzerland, 2009.

31. Noghabi, A.S.; Sadeh, J.; Mashhadi, H.R. Considering different network topologies in optimal overcurrent relay coordination using a hybrid GA. *IEEE Trans. Power Deliv.* **2009**, *24*, 1857–1863. [CrossRef]

32. Tjahjono, A.; Anggriawan, D.O.; Faizin, A.K.; Priyadi, A.; Pujiantara, M.; Taufik, T.; Purnomo, M.H. Adaptive modified firefly algorithm for optimal coordination of over-current relays. *IET Gener. Transm. Distrib.* **2017**, *11*, 2575–2585. [CrossRef]

33. Dahej, A.E.; Esmaeili, S.; Hojabri, H. Co-Optimization of protection coordination and power quality in microgrids using unidirectional fault current limiters. *IEEE Trans. Smart Grid* **2018**, *9*, 5080–5091. [CrossRef]

34. Sharaf, H.M.; Zeineldin, H.H.; Ibrahim, D.K. A proposed coordination strategy for meshed distribution systems with DG considering user-defined characteristics of directional in-verse time over-current relays. *Int. J. Electr. Power Energy Syst.* **2015**, *65*, 49–58. [CrossRef]

35. Jiang, X.; Li, S. BAS: Beetle antennae search algorithm for optimization problems. *arXiv* **2017**, arXiv:1710.10724. [CrossRef]

Publisher's Note: MDPI stays neutral with regard to jurisdictional claims in published maps and institutional affiliations.

Article

Modelling and Fault Current Characterization of Superconducting Cable with High Temperature Superconducting Windings and Copper Stabilizer Layer

Eleni Tsotsopoulou [1,*], Adam Dyśko [1], Qiteng Hong [1], Abdelrahman Elwakeel [1],
Mariam Elshiekh [1,2], Weijia Yuan [1], Campbell Booth [1] and Dimitrios Tzelepis [1]

[1] Department of Electronic and Electrical Engineering, University of Strathclyde, Glasgow G1 1XW, UK;
a.dysko@strath.ac.uk (A.D.); q.hong@strath.ac.uk (Q.H.); abdelrahman.elwakeel@strath.ac.uk (A.E.);
mariam.elshiekh@strath.ac.uk or mariam_elshaikh@f-eng.tanta.edu.eg (M.E.);
weijia.yuan@strath.ac.uk (W.Y.); campbell.d.booth@strath.ac.uk (C.B.); dimitrios.tzelepis@strath.ac.uk (D.T.)

[2] Department of Electronic and Electrical Engineering, Tanta University, Tanta 31512, Egypt

* Correspondence: eleni.tsotsopoulou.2018@uni.strath.ac.uk

Received: 22 November 2020; Accepted: 15 December 2020; Published: 16 December 2020

Abstract: With the high penetration of Renewable Energy Sources (RES) in power systems, the short-circuit levels have changed, creating the requirement for altering or upgrading the existing switchgear and protection schemes. In addition, the continuous increase in power (accounting both for generation and demand) has imposed, in some cases, the need for the reinforcement of existing power system assets such as feeders, transformers, and other substation equipment. To overcome these challenges, the development of superconducting devices with fault current limiting capabilities in power system applications has been proposed as a promising solution. This paper presents a power system fault analysis exercise in networks integrating Superconducting Cables (SCs). This studies utilized a validated model of SCs with second generation High Temperature Superconducting tapes (2G HTS tapes) and a parallel-connected copper stabilizer layer. The performance of the SCs during fault conditions has been tested in networks integrating both synchronous and converter-connected generation. During fault conditions, the utilization of the stabilizer layer provides an alternative path for transient fault currents, and therefore reduces heat generation and assists with the protection of the cable. The effect of the quenching phenomenon and the fault current limitation is analyzed from the perspective of both steady state and transient fault analysis. This paper also provides meaningful insights into SCs, with respect to fault current limiting features, and presents the challenges associated with the impact of SCs on power systems protection.

Keywords: superconducting cable; quench; high temperature; coppers stabilizer; superconducting tape; fault current limiting feature

---

1. Introduction

Transmission System Operators (TSOs) are responsible for the security of power grids and maintaining the balance between power generation and demand. However, new trends have emerged in power systems, pushing for a change in the way that networks are controlled, giving the TSOs plenty of new challenges to face in order to maintain the reliability and the security of power exchanges. The traditional power grids are gradually evolving towards power networks with high penetration of large-scale Renewable Energy Sources (RES) in both distribution and transmission level. In addition, more of the networks' equipment are reaching their capacity limits, while at the same time the utilities face several converging challenges caused by demand growth. All these factors bring about new challenges for future power systems, requiring the development of bulk power corridors as interconnections between different countries, and the upgrading of existing networks. Consequently, in order to avoid technologically, economically, and socially challenging solutions, such as building of new substations [1], there is a need for the investigation of new technologies which can overcome these restrictions and increase the electrical capacity and flexibility of the network.

In addition, the penetration of RES changes significantly the fault levels and the resulting fault current signatures. Such changes imply the need for upgrading the existing switchgear and protection systems. As a result, the utilization of Resistive Superconducting Fault Current Limiters (RSFCLs) has been proposed by [2–5] as a viable solution towards addressing the challenge of managing short circuit currents in power-dense systems. However, the RSFCLs are very expensive. Therefore, many researchers have been focused on the integration of fault current limiting features into other power system devices, in an attempt to take advantage of the unique features of the superconducting materials while fulfilling the cost requirements [6]. The studies presented in this paper promote that the utilization of SCs with a copper stabilizer layer connected in parallel. The main scope of this research is to study the fault current and voltage signatures resulted by the utilization of the SCs. Emphasis is given to the fault current limitation feature (as an extension to the cable's primary function as a lossless transmission media during steady state operation), in conjunction with the assessment of the potential benefits of the copper stabilizer layer during transient phenomena. The obtained results provide useful information regarding the fault analysis of future power grids integrating SCs and high amounts of RES, which can be considered as a prerequisite step for designing effective protection schemes.

1.1. Characteristics of Superconducting Cables

In recent years, the deployment of Superconducting Cables (SCs) in power system applications has become widely accepted due to their unique characteristics. Several prototype projects have been carried out worldwide which proposed the utilization of different configurations of SCs as a viable solution for bulk power transmission [7–12].

Compared to conventional cables, SCs are characterized by a plethora of technically-attractive features, such as higher current-carrying capability [13], higher power transfer at lower operating voltages and over longer distances [1,14], lower losses due to their lower resistance compared to that of overhead transmission lines [15], and more compact size due to their high current density. Therefore, the installation of SCs is considered a promising solution against congestion, especially in high power density areas such as metropolitan meshed networks. Furthermore, their fundamental property of transferring power over long distances, at low voltage levels, renders them the most effective way to interconnect renewable energy sources, such as offshore wind farms, to the power grid.

The superconducting behavior appears after cooling down the superconductor below a characteristic temperature, known as critical temperature T_C, which has a specific value for each superconducting material [16–19]. The maximum value of the current that can be conducted through the superconductor without encountering increase in the resistance value is called critical current I_C. However, superconductors lose their superconductivity if the magnetic field reaches its critical value H_C or in case the temperature increases beyond T_C. This phenomenon is called quench. These remarkable physical properties of SCs make them capable of conducting currents with approximately zero electrical resistance during steady state, while their variable resistance, which is dependent on the load current, in conjunction with the introduction of a high resistive layer into the superconducting wire, such as copper, result in fault current limitation in short-circuit situations. The contribution of SCs to fault current limitation is determined by the design.

1.2. Challenges Associated with the High Temperature Superconducting Cables Installation and the Superconducting Cables (SCs)

The discovery of HTS materials created the opportunity of applying the superconductivity principles to electric power devices such as, superconducting machines and SCs. The major advantage of the HTS materials is that their high critical temperature values, T_C, are attainable using liquid nitrogen, LN_2, as coolant with a boiling temperature of 77 K [20–22]. For the presented case studies, the Yttrium Barium Copper Oxide (YBCO) material has been chosen with $T_{C(YBCO)} = 93$ K, which belongs to the 2nd generation of HTS tapes (2G), as its transition from the superconducting state to normal state lasts for a few milliseconds, which makes it attractive considering the fault current limitation capability [8].

In addition, one of the most challenging tasks to be achieved is the connection between HTS cables and existing conventional circuits [23–25]. It is important to understand that the direction and the magnitude of power flows could be affected by the installation of HTS cables, due to their low impedance. During steady state conditions, HTS cables operate at the superconducting state, presenting the current path with approximately zero resistance and as a consequent, attracting naturally the power flow. These significant changes in the current distribution and the rearrangements of power flows must be considered in order to maintain power system stability.

Furthermore, the installation of HTS power cables impacts on the short-circuit level of the power system. The changes in the short-circuit level, and as a consequence the changes in the fault currents, affect the performance and design of power system protection schemes. The incorporation of the copper parallel layer and fault current limiting features in SCs cables have made them increasingly appealing for power system applications [26,27]. In steady state condition SCs transmit bulk power with low losses. Under fault conditions, when the fault current flowing through the HTS tapes exceeds the critical current I_C, the superconducting tapes will automatically quench and switch to normal resistive state. As the fault current increases, the resistance and the temperature of the cable increase as well, as interdependent variables. The transition from the superconducting state to the normal resistive state during short circuit conditions can occur within milliseconds (i.e., within a single AC cycle). Consequently, the integration of the fault current limiting feature into the HTS cable can limit the short-circuit current to a certain point, helping towards protecting the system [27]. This property of the SCs creates new challenges for the power system protection, as the calculation of the expected short-circuit level must be conducted in accordance with the variable resistance of the installed SCs.

The paper is organized as follows: Section 2 presents the detailed mathematical development of the utilized cable based on well-known equations which explain the behavior of superconductors. The model is developed using Matlab and Simulink software and is applied to a power system which contains wind farms and synchronous generators. In Section 3, different fault scenarios are carried out which aim to investigate the cable performance during transients and verify the practical feasibility of the proposed SCs model.

2. Modelling of SCs with 2G HTS Wires

Various numerical models of HTS cables have been recently proposed, which use the Finite Element Method (FEM) or finite-difference time-domain (FDTD) analysis to understand the non-linear electromagnetic properties of the superconductors [28–31]. The investigation of the electromagnetic and thermal properties of the HTS cables is an effective way to predict and optimize the cable performance under different operating conditions. However, for power system studies such as fault analysis, the performance of the numerical models using FEM and FDTD is compromised due to the computational complexity [30]. Thus, a simplified time-dependent model of a multilayer HTS cable will be analyzed in this research, providing a solid foundation for the utilization of SCs in power system applications.

2.1. Configuration and Design Specifications

Several design topologies of SCs have been developed to minimize the capital and operating costs. The different configurations can be classified based on the superconducting layer layout for each phase and the voltage level. One design, known as triaxial configuration involves three different phases attached onto a single former, contained in a single cryostat [1] as shown in Figure 1. The three phases are separated by a dielectric layer which provides electric insulation. The circulating liquid nitrogen flows between the copper screen and the inner cryostat wall to cool down the entire cable to a temperature range of 65–77 K [31]. This configuration offers higher carrying current capacity, and has the lowest inductance compared to other cable designs. Regarding the position of the insulation layer, SCs can be separated into two categories, namely the warm dielectric (WD) and the cold dielectric (CD), with the latter to be the most preferred design due to low losses and higher current capacity [32]. In this paper, a CD triaxial SCs with YBCO wires has been modelled. The detailed structure of the SCs tape is demonstrated in Figure 2.

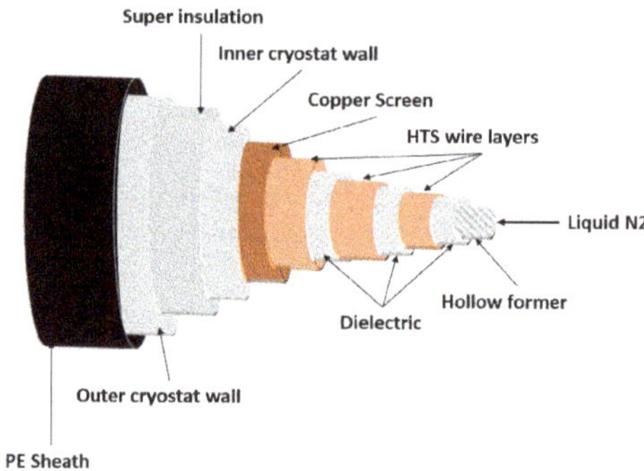

Figure 1. Configuration of triaxial SCs cable.

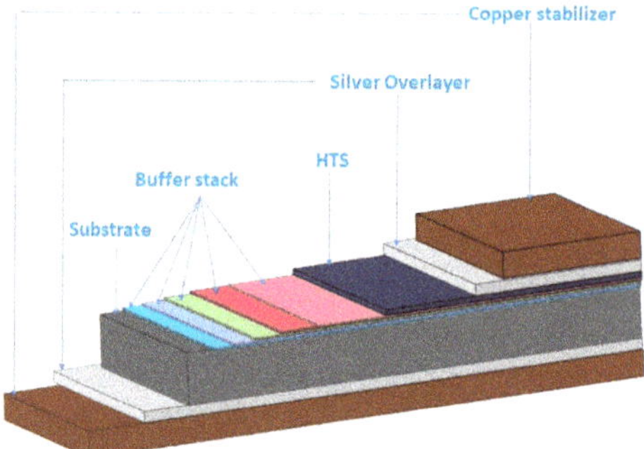

Figure 2. Configuration of SCs tape.

The typical structure of the YBCO tape consists of the YBCO layer, the copper stabilizer layer, the silver stabilizer layer, the Hastelloy substrate and the buffer layer which is placed between the substrate and the YBCO layer [33]. The YBCO layer, which is the only layer responsible for conducting the load current during the steady state operation, is manufactured as a film with very small thickness, protected by copper stabilizer layers on both sides. In the superconducting wire, a stabilizer layer (such as copper) is connected in parallel with the HTS layer to maintain stability, reduce the heat generation and the temperature during high current faults, and protect the cable from thermal-induced damage. This technique has been introduced and adopted by major manufacturers [34–37]. For the fault analysis, due to the parallel structure of the layers, the total fault current flowing through each phase must meet Equation (1),

$$I_{total} = I_{HTS} + I_{Copper} \tag{1}$$

where I_{total} is the total current, I_{HTS} is the current in YBCO layer and I_{Copper} is the current flowing in the copper layer. Specifically, as it is illustrated in Figure 3, in steady state, during which the HTS tapes are in the superconducting state, the load current only flows through the HTS layer (i.e., as presented by Equation (2)), due to its very low impedance compared to that of the copper stabilizer,

$$I_{total} = I_{HTS} \tag{2}$$

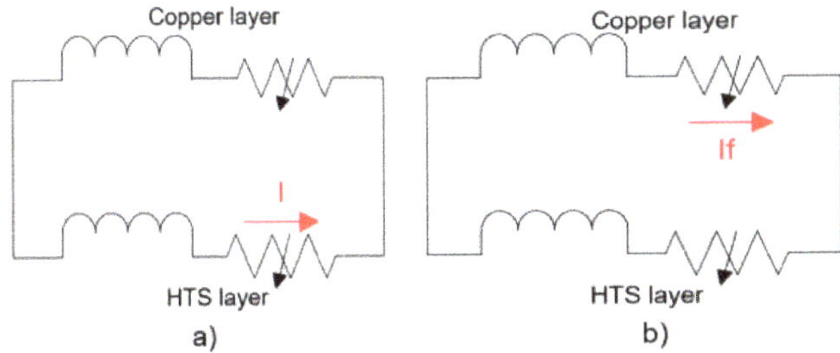

Figure 3. Operation of HTS cable during (**a**) steady state (**b**) fault.

In this case, during the steady state, the I_{Copper} is approximately zero.

In transient conditions, once the fault current exceeds the value of the critical current I_C the HTS tapes quench and their resistivity increases exponentially. Furthermore, the temperature of the HTS tapes is affected by the generated heat. The temperature increases gradually and exceeds the value of the critical temperature T_C, indicating the transition to the normal state. Once the HTS tapes enter the normal state, the variable resistance, which is a function of the current density J and the temperature T, reaches values which are much higher than that of the copper layer. Hence, the transient current is diverted into the copper stabilizer layer, as expressed in Equation (3), which acts as a by-pass circuit. Thus, the effect of the stabilizer layer is important for the transient studies,

$$I_{total} = I_{Copper} \tag{3}$$

where I_{Copper} is the diverted fault current, flowing through the stabilizer layers, while a very small current (approximately zero) flowing through the HTS layers.

Based on the analysis presented above and according to the study conducted in [37], the boundary of the critical current I_C determines whether or not the superconducting tape quenches. Thus, exceeding the threshold of I_C can be considered as the impelling factor that leads to quench, while the threshold of the critical temperature T_C determines if the superconductor will enter the highly resistive normal state. Therefore, it can be defined as a criterion for the degree of quenching. To further study the performance of the integrated HTS cables, it is of major importance to investigate in more detail the transition period from the superconducting to the normal state. To study the quenching process, special focus should be given to the current distribution among the layers and the resistance variations with respect to the accumulated heat and the current amplitude. In the following part, the proper design of a simplified model of multilayer HTS power cable will be presented.

2.2. Development of SCs Model

2.2.1. Equivalent Circuit

Each phase of the cable consists of (i) several HTS tapes connected in parallel, in order to cope with the large operating current, and (ii) two copper-stabilizer layers connected in parallel with the HTS layer. The rest of the cable layers shown in Figure 2 have been neglected for simplicity reasons as the increase of the temperature mainly affects the resistance of the HTS and the copper stabilizer layer. The number of the tapes and the layers have been selected after taking into consideration the value of the designed critical current I_C, while the geometric characteristics of the tapes have been determined based on the maximum quenching voltage [38]. In particular, the rated current I_{rated} during the steady state operation has been considered equal to 80% of the critical current I_C [39]. Therefore, the number of tapes can be calculated by the following equation,

$$I_{rated} = 0.8 \cdot I_{C\_initial\_per\_tape} \cdot n \tag{4}$$

where $I_{C\_initial\_per\_tape}$, corresponds to the initial value of the critical current for each YBCO tape, and has been estimated based on validated manufacturers' data presented in [8], where n is the number of tapes.

The equivalent impedance of each phase is dependent on the current distribution among the HTS and the copper layers. Figure 4. shows the equivalent circuit of the three phase triaxial SCs. The resistance of the HTS layers is introduced as a variable resistance which represents the quench phenomenon with an initial value of approximately zero. The PI section model has also been used, in order to implement the self- and mutual-inductances and the capacitance of the cable. The resistance of the copper stabilizer has been modelled as a variable resistor. Once the current increases to higher than the critical value I_C, the HTS tapes resistance starts to increase and the current flows in both the superconducting and the copper layers. During this process, heat is generated in the tape resulting in a dramatic temperature rise. Once the temperature exceeds T_C , the cable reaches the normal state mode and the current flows through the copper layer.

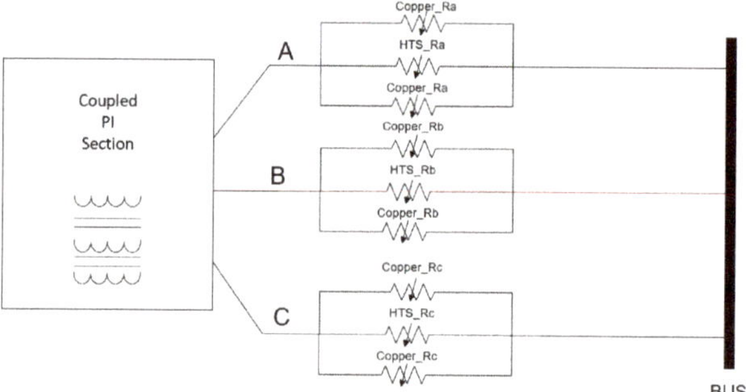

Figure 4. Equivalent electrical circuit of the modeled cable.

2.2.2. Modelling Methodology

The following Section presents the detailed equations that have been used for the SCs development and the subprocess that has been followed to calculate the resistance of each layer. The modelling method followed is based on the equations proposed by the authors in [40], in which the modelling of transformers with superconducting windings is presented. To describe the HTS and copper layers, the operation of the multilayer SCs has been divided into three modes (referred as three distinct stages for simplicity) with respect to the current distribution and the values of the equivalent resistance. Stage 1 refers to the superconducting mode, where the applied current is lower than the critical current I_C and the temperature is considered to be below the critical temperature T_C,

$$I_{applied} < I_C \tag{5}$$

$$T = T_{operating} \tag{6}$$

where $T_{operating}$ is the operating temperature of 70 K.

Stage 2 refers to the flux flow mode, when the quench starts, and is determined by the following boundary conditions:

$$I_{applied} > I_C \tag{7}$$

$$T < T_C \tag{8}$$

At this stage the HTS tapes start to quench and their resistivity increases sharply as a function of the current density J_C and the accumulated heat.

At the final mode, stage 3, which is described by the boundary conditions (9) and (10), the HTS layer completely loses its superconductivity and enters normal state.

$$I_{applied} > I_C \tag{9}$$

$$T > T_C \tag{10}$$

The main parameters that affect the resistance value and the operation mode of the HTS tapes-layers are the critical current density J_C, and the critical temperature T_C [7]. The relationship between the temperature T, the current density J_C and the critical current I_C is given by the following equations,

$$J_C(T) = \begin{cases} J_{C0} \cdot (\frac{(T_C - T(t))^a}{(T_C - T_0)^a} \\ 0, \; T > T_C \end{cases} \tag{11}$$

$$J_{C0} = \frac{I_{C\_initial}}{s_{HTS}} \tag{12}$$

$$I_{C\_initial} = 267 \cdot n \tag{13}$$

$$s_{HTS} = w_{HTS} \cdot t_{HTS} \cdot n \tag{14}$$

where J_{C0} is the critical current density (A/m^2) at the initial operating temperature $T_0 = 70$ K; $T_C = 92$ K is the critical temperature of the HTS superconducting tape; the density exponent a is 1.5; $I_{C\_initial}$ corresponds to the initial value of the critical current and s_{HTS} is the cross section area of the superconductor; w_{HTS} is the width of the HTS material and t_{HTS} is the thickness of the HTS material and n is the number of tapes. As can be seen from Equations (11)–(13), the value of the critical current density $J_C(T)$, and by extent the value of the critical current $I_{C\_initial}$, decreases drastically as the temperature $T(t)$ rises. The temperature dependence on the critical current density is known in literature as 'critical current density degradation' [41]. The effect of the resulting degradation must be taken into consideration for the design of large-current-capacity AC SCs and their cooling systems.

To better understand the operation of the HTS cable it is crucial to estimate the resistance of the HTS and the copper stabilizer layers and the equivalent resistance of the SCs at every stage. Initially, at stage 1, the HTS tape is in a superconducting state. The resistivity of the HTS tape is $\rho_0 = 0$ ($\Omega \cdot$m) and therefore its total resistance equals approximately zero. The copper stabilizer resistance has been considered constant and the total equivalent resistance of the cable is equal to the HTS layer resistance, as the main current flows through it only. At stage 2, when the applied current exceeds the value of the critical current, the resistivity of the HTS tape increases exponentially as a function of the current density and the temperature, according to Equation (15),

$$\rho_{HTS} = \frac{E_C}{J_C(T)} \cdot \left(\frac{J}{J_C(T)} \right)^{N-1}, \; I > I_C, \; T < T_C \tag{15}$$

where $E_C = 1$ μV/cm is the critical electric field; the coefficient N has been selected to be 25, while the YBCO tapes should be within the range of 21 to 30 [42]. The copper stabilizer resistance corresponds to a constant value, similar to that of stage 1. This approximation can be confirmed by the small variation of copper resistivity with the temperature rise at this stage. The total resistance of the superconductor is obtained by the equation for equivalent resistance of parallel electrical circuits,

$$R_{SC} = \frac{R_{HTS} \cdot R_{Cu}}{R_{HTS} + R_{Cu}} \tag{16}$$

where R_{SC} is the total resistance of the SC during stage 2.

When $T > T_C$, stage 3 has been initiated, which corresponds to the normal resistive mode. The HTS layer-tape resistance reaches values much larger than the copper stabilizer resistance. For modelling purposes, a maximum limit has been set for the HTS resistance value at stage 3. However, in this case the resistivity (Ω·m) of the copper changes with respect to temperature rise and is determined by Equation (17). The maximum value that copper resistivity can reach is calculated for $T = 250$ K, which has been selected as the upper temperature limit [43].

$$\rho_{Cu} = (0.0084 \cdot T - 0.4603) \cdot 10^{-8}, 250 \text{ K} \geq T > T_C \tag{17}$$

During the normal mode, the equivalent resistance of the SFCLC is affected solely by the value of the copper stabilizer resistance, as the transient current is diverted into the copper layers.

2.2.3. Thermal Transfer Analysis during the Quenching

Superconducting tapes are immersed in liquid nitrogen LN_2, which is used as a refrigerant for cooling the SCs below a certain temperature. When the resistance of the HTS tapes is zero, (stage 1) the amount of the power dissipated is not considered significant. When a fault occurs, the resistance increases, and heat is generated by the superconductor. The generated heat increases the superconductor temperature and part of it is absorbed by the LN_2 circulation system (the heat transfer with the external environment has been neglected). The power dissipated is a function of the fault current and can be calculated by Equation (18),

$$P_{diss} = i(t)^2 \cdot R_{SC} \tag{18}$$

where t is time and R_{SC} is the equivalent resistance of the superconductor.

The cooling power that can be removed by the LN_2 cooler is given by Equation (19),

$$P_{cooling} = h \cdot A \cdot (T(t) - 70) \tag{19}$$

where $T(t)$ is the temperature; A is the total area that is covered by the cooler; h is the heat transfer coefficient. The heat transfer coefficient is a function of the temperature and considered as the major factor which determines the cooling system effectiveness and the cable recovery, representing the heat transfer process between the superconducting tapes and the LN_2. Equations (20)–(23) below present the calculation of h based on the temperature variation [44].

$$h = 125 + 0.069 \cdot \Delta T, \ 56.3 \leq \Delta T \leq 214 \tag{20}$$

$$h = 12292.13 - 709.32 \cdot \Delta T + 14.735 \cdot \Delta T^2, \ 18.94 \leq \Delta T \leq 56.3 \tag{21}$$

$$h = 82.74 - 131.22 \cdot \Delta T + 37.64 \cdot \Delta T^2, \ 4 \leq \Delta T \leq 18.94 \tag{22}$$

$$h = 21.945 \cdot \Delta T, \ 0 \leq \Delta T \leq 4 \tag{23}$$

If Equation (19) is subtracted from Equation (18) then the net power P_{SC} can be calculated. Equation (24) is the thermal equilibrium equation which gives the part of the dissipated power which leads to temperature rise in the superconductor during the quenching process.

$$P_{SC}(t) = P_{diss}(t) - P_{cooling}(t) \tag{24}$$

Finally, Equation (25) gives the temperature $T(t)$ of the superconducting tapes at each iteration step,

$$T(t) = T_0 + \frac{1}{C_p} \cdot \int_0^t P_{SC}(t)dt \tag{25}$$

where T_0 is the initial temperature of the HTS materials and C_p (J/K) is the heat capacity.

For stage 2, when the quenching starts, the current starts to flow through the copper layer. However, as the temperature rise is not very high at this stage, the copper heat capacity variation with the temperature is neglected. The heat capacity of the YBCO material can be calculated by Equation (26) and the volume of the cable by Equation (27),

$$C_P = 2 \cdot T \cdot d \cdot v \qquad (26)$$

$$v = l \cdot th \cdot w \cdot n \qquad (27)$$

where d is the density of the material, T is the temperature, v is volume and l is length; th is the thickness and w is the width and n is the number of tapes. At stage 3, when the resistance of the HTS tapes-layer has reached very high values due to the increased temperature, the fault current flows through the copper layers. In this case the heat capacity in Equation (28) is substituted by the total heat capacity of the superconductor Equation (29) gives the heat capacity of the copper layer,

$$C_P = C_{PHTS} + C_{Cu} \qquad (28)$$

$$C_{PCu} = C_{Cu} \cdot d_{Cu} \cdot v \qquad (29)$$

where C_{Cu} is the heat capacity of the copper and d_{Cu} is the density of the copper.

The classification of the quenching process and the corresponding characteristics of each stage are listed in Table 1.

Table 1. Quenching characteristics.

	Quenching Process		
Parameters	**1. Supercondu-Cting State**	**2. Quenching: Flux Flow State**	**3. Quenching: Normal State**
Critical current/temperature	$I_{applied} < I_C$	$I_{fault} > I_C,\ T < T_C$	$I_{fault} > I_C.\ T > T_C$
Resistivity of HTS layer ρ_{HTS}	$\rho_{HTS} = 0$	$\rho_{HTS} = \frac{E_c}{Jc(T)} \cdot \left(\frac{J}{Jc(T)}\right)^{N-1}$	$\rho_{HTS} = max\_value$
Resistivity of Copper layer ρ_{Cu}	$\rho_{Cu} = \rho_{Constant}$	$\rho_{Cu} = \rho_{Constant}$	$\rho_{Cu} = (0.0084 \cdot T - 0.4603) \cdot 10^{-8}$
Current distribution	$I_{applied} > I_{HTS}$	$I_{HTS} > I_{Cu}$	$I_{Cu} > I_{HTS},\ I_{fault} = I_{Cu}$

Matlab has been used to model Equations (11)–(29) in order to compute the resistance values of the HTS tapes R_{HTS}, the copper stabilizer R_{Cu}, and the variation of the temperature ΔT of the superconductor. The calculation process is shown in Figure 5. $T_{o,t}$, $I_{C\_initial}$, $J_{Cinitial}$ are the initial values of the operating temperature, critical current and critical current density for the first iteration, respectively. Once the I_{rms} gives a value of current density J_i, which exceeds the critical value J_C, the HTS tapes start to quench. During the quenching process the values of I_{rms} P_{diss}, $P_{cooling}$, P_{SC}, and T_{i+1} are updated in each time step, T_{step}. The calculation process terminates once the T_{i+1}, R_{HTS}, and R_{Cu} reach their maximum values, indicating that superconductor has entered into normal state.

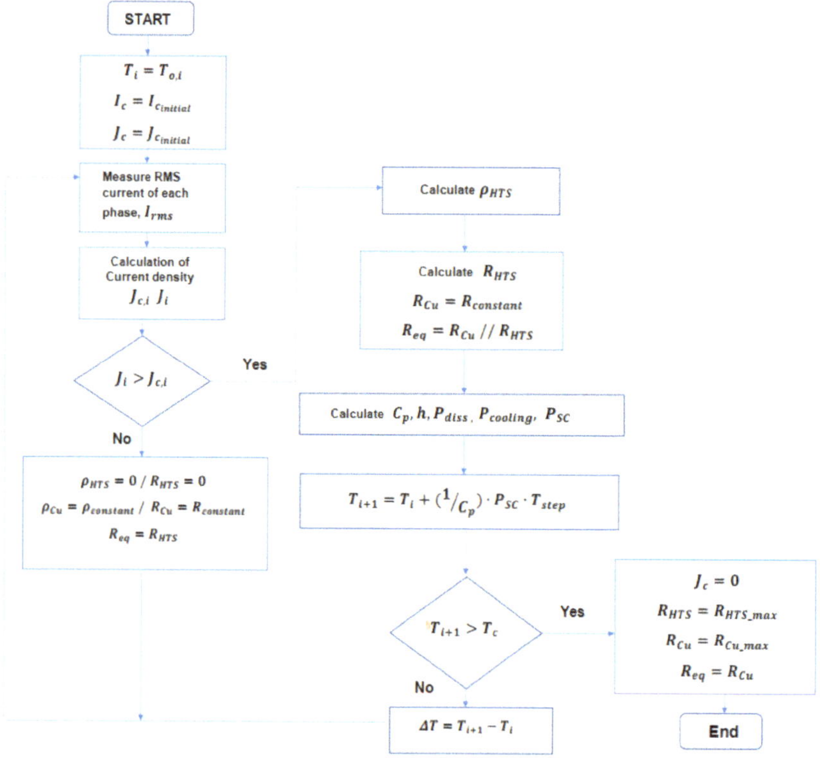

Figure 5. Flowchart corresponding to the calculation process for the resistance values of the HTS tapes R_{HTS}, the copper stabilizer R_{Cu}, the equivalent R_{eq}, and the variation of the temperature ΔT of the superconductor.

3. Simulation Results

In this Section the model development is completed by integrating a lumped model of a 5 km long SCs into a simulated power system which contains converter-connected generators and a synchronous generator (SG). The fault analysis is carried out, analyzing the stages of the quenching process, and the corresponding plots of the fault current signatures, resistance values, and temperature have been obtained. For the purpose of the simulation-based fault analysis, the system under test (as shown in Figure 6) has been built in Matlab and Simulink shows the components of the tested system. Table 2 presents the main components of the power system.

The network consists of an equivalent voltage source connected at Bus 1 with a nominal voltage of 275 kV, which represent the equivalent connected transmission system. Two different generation units accounting for (i) a wind farm connected via Voltage Source Converter (VSC) and (ii) a Synchronous Generator (SG) are connected at Bus 11. The SG has been modelled as a standard salient pole synchronous machine with an automatic voltage regulator (AVR) and a power system stabilizer. The wind farm consists of 100 variable speed wind turbines, which consist of permanent magnet SGs connected via VSC and operate under a Direct Quadrature Current Injection (DQCI) control algorithm. The 132 kV/10 km transmission lines transfer power to (132 kV/33 kV) transformers. The 33 kV triaxial SCs connects Bus 7 and Bus 11, and due to its high-power density it is capable of transfering power up to 202 MVA. For the steady state, the resistance of the HTS tapes has been considered approximately zero while the positive and zero sequence inductance and capacitance have been obtained by [23]. Regarding the final stage of the quenching process, known as normal state, for simulation purposes, a maximum value has been set for both the HTS and copper stabilizer layers. The idea behind this assumption was to model the change of HTS and copper layer resistance according to the current and temperature changes and examine the current distribution among the different layers during the quench phenomenon. These assumptions can be considered reasonable as the HTS layers become highly resistive during the fault which results in the flow of short-circuit current through the stabilizer. Therefore, for the normal state, a high resistance value has been selected for the HTS layer, based on the studies conducted in [40]. The maximum resistivity of the copper stabilizer layer has been calculated by using Equation (17) for a temperature $T = 250$ K. The corresponding parameters and the specifications of the proposed HTS cable are listed in Tables 3 and 4.

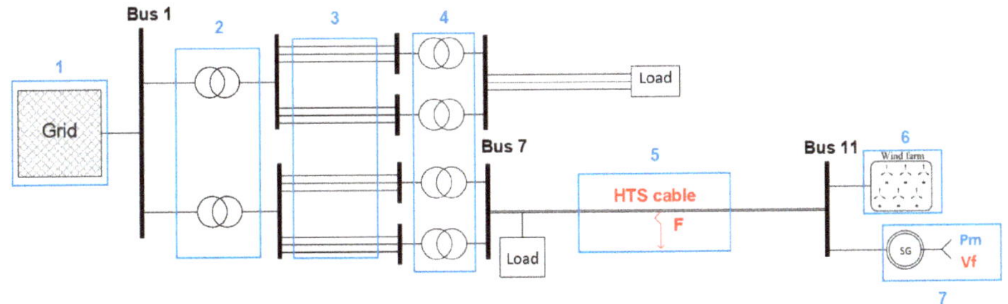

Figure 6. Case study test network.

Table 2. Components of the tested system.

Number	Components
1	Grid
2	Transformers 275 kV/132 kV, 150 MVA
3	10 km transmission lines
4	Transformers 132 kV/33 kV
5	5 km SC
6	Converter-interfaced generator (Wind farm)
7	Synchronous generator

Table 3. SC Parameters.

Cable Parameters		
Parameters	**Symbol**	**Value**
Cable length	l	5 km
No. of YBCO tapes	n	15
Critical current	I_C	$256 \cdot n$
Critical temperature	T_C	92 K
Width of YBCO tape	w	0.004 m
Thickness of YBCO tape	t_{HTS}	1 μm
Thickness of copper tape	t_{Cu}	40 μm
Density of YBCO tape	d_{HTS}	5900 (kg/m³)
Density of copper tape	d_{Cu}	8940 (kg/m³)

Table 4. Specifications of 33 kV SC.

Cable Specifications		
Parameters	**Symbol**	**Value**
Rated Voltage	V	33 kV
Operating Temperature	T	70 K
Rated Capacity	S	202 MVA

In the following part, systematic iterative simulations have been performed, which include (i) 3-Phase-to-ground faults at two different fault locations, (ii) a Phase-to-Phase-to-ground fault and (iii) a Phase-to-ground fault. In all cases the faults initiate at $t = 5.06$ s and last for 120 ms. To obtain a high-fidelity insight of the transient phenomena of SCs, a sampling frequency of $f = 2$ MHz has been used (accounting for simulations and records).

3.1. Fault Analysis of the SCs

Initially, a 3-Phase-to-ground fault with fault resistance $R_f = 0.01\ \Omega$ was triggered at 50% of the HTS cable's length at $t = 5.06$ s, and it was cleared after 120 ms. Figure 7 shows the stages of the quenching process, Figure 8a illustrates the fault current signatures contributed by the wind farm and the SG at Bus 11, while Figure 8b presents the corresponding voltage signatures. The resulting fault current distribution among the different layers of the three phases is shown in Figure 8c–h. At the superconducting state (stage 1) and the flux flow state, which is a moderately resistive state, the current flows through the HTS layers, presenting high peaks due to the low resistance of the superconductor when the fault occurs. However, as the fault current exceeds the critical value I_C in the flux flow mode (stage 2), the temperature rises continuously and the value of the resistance of the superconductor increases rapidly to very high values, reaching the normal state (stage 3). Therefore, as it can be seen from Figure 8d,f,h, the main current has been diverted to the copper stabilizer layers, indicating that the normal state has been reached, while the HTS layers conduct approximately zero current. Figures 9 and 10 illustrate the changes in the resistance values and the temperature rise, respectively. Initially, the temperature is 70 K for the three phases and the equivalent resistance of the superconductor is approximately zero. Once the temperature exceeds 92 K, which is the critical value, at $t = 5.064$ s, the HTS tapes enter the normal state and their resistance starts to increase rapidly. For stage 2, the equivalent resistance of the superconductor is calculated based on Equation (16). The current distribution starts to change, and the fault current is diverted to the stabilizer layers. Subsequently, in the normal state (stage 3) the equivalent resistivity is equal to the maximum resistivity of the copper stabilizer layer obtained by Equation (17). Therefore, the proposed design has achieved the current sharing between the HTS and stabilizer layers, aiming to improve the performance of the cable and self-protecting it from being destroyed.

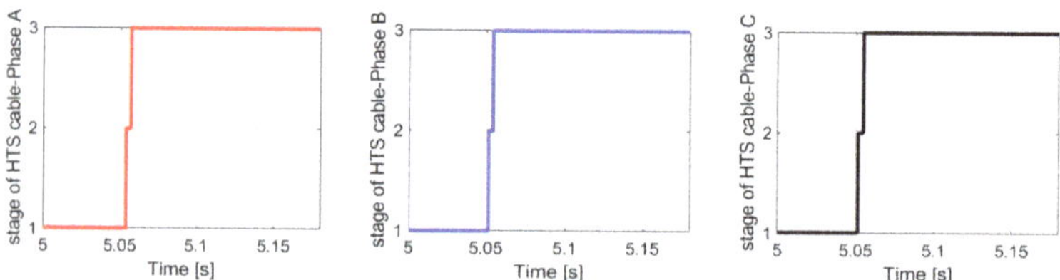

Figure 7. Stages of quenching process for phase A, B, and C for 3-Phase-to-ground solid fault at 50% of cable's length.

As discussed earlier, the installation of the SCs impacts the magnitude of fault currents. Indeed, from the fault current waveforms plotted in Figure 8, at the time of the fault event at $t = 5.06$ s, the highest first current peak is approximately 15 kA. As the value of the layers' resistance increases immediately, the magnitude of the fault currents decreases. Specifically, at $t = 5.064$ s, when the values of resistances and the temperature reach high values, the fault current starts flowing through the stabilizer layers, presenting peaks of approximately 5.5 kA. During the current elimination within the first fault cycle, some peaks are presented at the 3-Phase fault voltages. Moreover, it is noticeable that after $t = 5.069$ s and before the fault clearance at $t = 5.18$ s, the magnitude of fault currents at Bus 11 are limited and the phase voltages show higher magnitudes compared to steady state. This is interpreted based on the large equivalent resistance inserted by the SCs. Hence, it is evident that SCs provide effective limitation of fault currents in systems containing SGs and converter-interfaced generators. Such fault current limiting capability seems to be an interesting feature in regards towards protecting networks with varying short-circuit levels. Furthermore, the high voltage magnitudes during transient conditions raises new challenges for the voltage-assisted protection schemes. Normally, during the fault events, the voltage magnitude is anticipated to be reduced. However, in this case, when the fault occurs at $t = 5.06$ s, the 3-Phase voltages decrease for few milliseconds, but when the equivalent resistance of the superconductor increases, the fault current decreases, while the 3-Phase voltages present high peaks. The introduction of high equivalent resistance leads to voltage spikes across the superconductor. The faults at the SCs can be considered as high impedance faults in nature, jeopardizing the operation of the existing protection schemes.

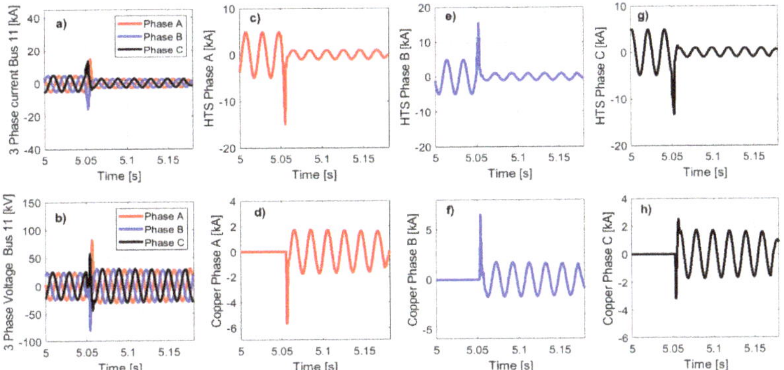

Figure 8. Fault current and voltage signatures for 3-Phase-to-ground solid fault at 50% of cable's length.: (**a**) phase currents at Bus 11, (**b**) phase voltages at Bus 11, (**c**) current in HTS layer of phase A, (**d**) current in copper layer of phase A, (**e**) current in HTS layer of phase B, (**f**) current in copper layer of phase B, (**g**) current in HTS layer of phase C, (**h**) current in copper layer of phase C.

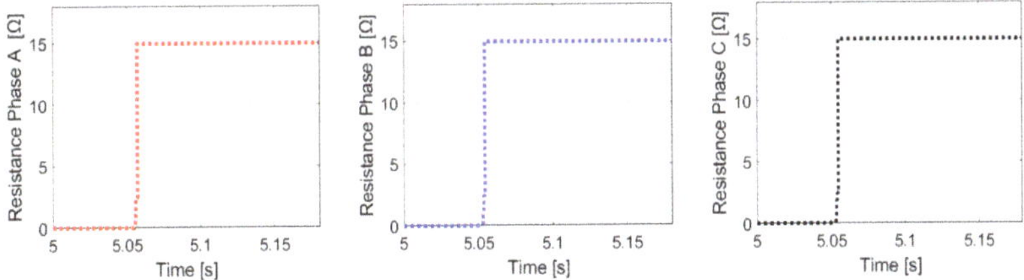

Figure 9. Equivalent resistance for phases A, B, and C for 3-Phase-to-ground solid fault at 50% of cable's length.

Additionally, the fault currents, the voltage signatures, and the current distribution characteristics for a Phase-A-to-ground and a Phase-A-B-to-ground faults at the 50% of the HTS cable's length with $R_f = 0.01$ Ω are reported in Figure 11 and Figure 12, respectively. The faulted phases of the proposed SCs have been found to behave in a similar way as in the previous case of the 3-Phase-to-ground fault. The characteristics of the superconductor resistance have the same trend as those presented in Figure 9 for the faulted phases. However, the equivalent resistance of the HTS layers of non-faulted phases remains at 0 Ω, as they do not quench and operate at superconducting state. Regarding the temperature rise for the faulted phases, it can be described based on Figure 10, while for the non-faulted phases the operating temperature remains constant at 70 K prior to and during the fault. For the non-faulted phases, the fault current flows only through the HTS layer. Therefore, the specific design target of the current limitation can be verified for different fault types with approximately zero fault resistance.

Once the value of the fault current density exceeds the critical value $J_C(T)$ (Equation (11)) within the first fault cycle, the resistivity of the HTS layer increases based (refer to Equation (15)) and the fault current diverts to the copper stabilizer layer. The feasibility of the parallel stabilizer layer can been confirmed for 3-Phase-to-ground, Phase-A-to-ground and Phase-A-B-to-ground faults by observing the current distribution characteristics in Figure 8, Figure 11, and Figure 12 respectively. As the quenching process evolves, the temperature of the SCs increases, reaching values higher than the critical T_C; during the normal resistive mode, the value of the SCs equivalent resistance is only determined by the value of the stabilizer layer given by Equation (17). The further increase in temperature results in an increase in the resistivity of the copper stabilizer layer (Equation (17)) which leads to further reduction in fault current. The accuracy of the fault current limiting capability is verified by Figure 8 for a 3-Phase-to-ground fault, where the first peak of the fault current at $t = 5.06$ s is approximately 15 kA; however, within the first cycle, and before the fault clearance at $t = 5.18$ s, the peak of the fault current is reduced to 1.8 kA. The same behaviour is observed for a Phase-A-to-ground and Phase-A-B-to-ground fault, as depicted by Figure 11 and Figure 12, respectively. The first peak of the fault current flowing through HTS layer, has values of 15 kA for the faulted phases, while the resulted fault current flowing through the stabilizer layer has been limited to approximately 1.7 kA.

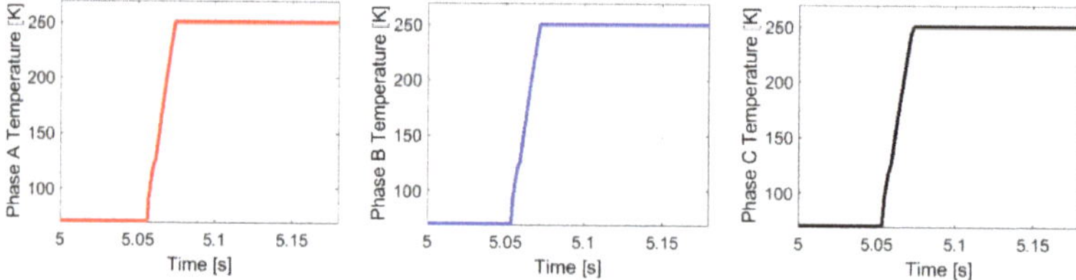

Figure 10. Temperature for phases A, B, and C for 3-Phase-to-ground fault at 50% of cable's length.

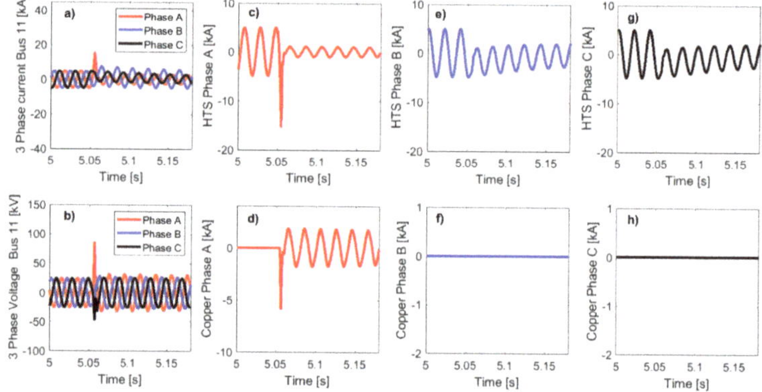

Figure 11. Fault current signatures for Phase-A-to-ground solid fault at 50% of cable's length: (**a**) phase currents at Bus 11, (**b**) phase voltages at Bus 11, (**c**) current in HTS layer of phase A, (**d**) current in copper layer of phase A, (**e**) current in HTS layer of phase B, (**f**) current in copper layer of phase B, (**g**) current in HTS layer of phase C, (**h**) current in copper layer of phase C.

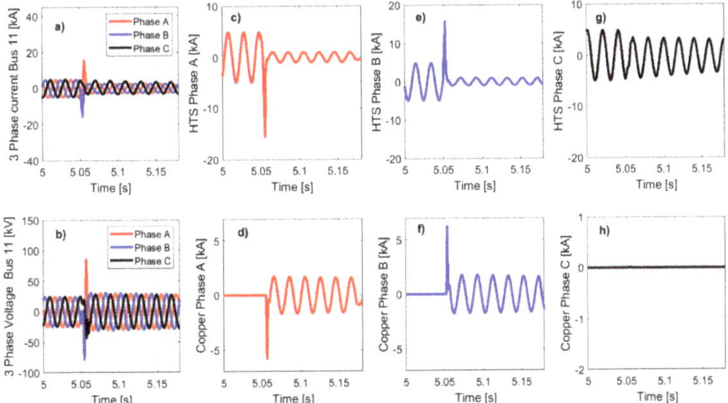

Figure 12. Fault current signatures for Phase-A-to-ground Phase-A-B-to-ground solid fault at 50% of cable's length: (**a**) phase currents at Bus 11, (**b**) phase voltages at Bus 11, (**c**) current in HTS layer of phase A, (**d**) current in copper layer of phase A, (**e**) current in HTS layer of phase B, (**f**) current in copper layer of phase B, (**g**) current in HTS layer of phase C, (**h**) current in copper layer of phase C.

3.2. Current Limitation

In this Section, the presented analysis aims to evaluate the transient performance of the SCs in contrast with a conventional copper cable installed at the same power system. For this reason, emphasis has been given on the calculation of the current-limitation capability as a percentage of the prospective fault current flowing through a conventional copper cable, during the quenching process. In particular, a 3-Phase-to-ground fault with fault resistance of $R_f = 0.01\ \Omega$ was applied at the 50% of the SCs length. The same fault has been repeated for the case of conventional copper cable. The fault currents captured by the SCs model during the simulations have been compared with the prospective fault currents through the conventional copper cable, highlighting the merits arising by utilizing superconductors. Figure 13 demonstrates the RMS value of the fault currents at Bus 11 during a 6-cycle 3-Phase-to-ground fault for both cases.

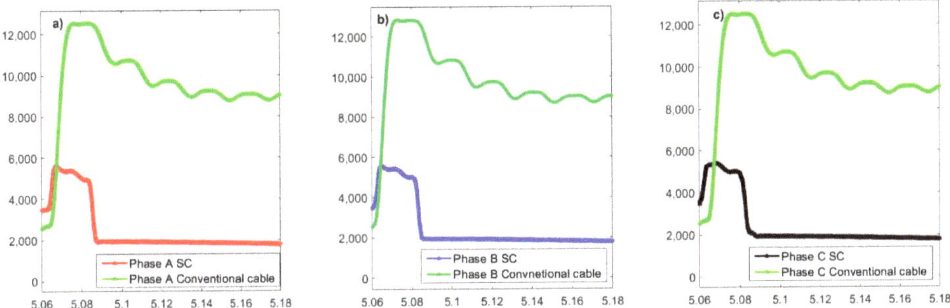

Figure 13. RMS values of the fault currents at Bus 11 during 3-Phase-to-ground solid fault at 50% of the proposed SCs and a conventional copper cable: (**a**) Phase A, (**b**) Phase B, (**c**) Phase C.

Similarly, to the previous Section, the fault is initiated at $t = 5.06$ s and cleared after 120 ms. When the fault occurs at $t = 5.06$ s, the RMS value of the current for the SCs is slightly higher compared to that of the conventional cable, as at the initial quenching state (stage 2) the resistance of the HTS tapes has not reached high values yet. It is well-established that the short-circuit magnitude is determined by the X/R ratio of the circuit. Therefore, it can be seen that the RMS values of the fault currents start to decrease at the time instant of $t = 5.065$ s, due to the high resistance and the significant temperature increase.

To quantify the fault current limitation by adding the fault current limiting function, a current limitation percentage of the prospective current through a conventional cable has been introduced based on Equation (30). Particularly, for the case of the SCs the RMS values of the limited fault currents during the whole quenching process (stage 2 and stage 3) have been calculated and compared with the prospective current values. Figure 14 shows the current limitation percentage per phase, verifying and supporting the practical feasibility of the proposed cable design,

$$I_{current-limitation}(\%) = \frac{I_{conv} - I_{SC}}{I_{conv}} \cdot (100\%) \tag{30}$$

where I_{conv} is the RMS value of the fault current flowing through the conventional copper cable and I_{SC} is the fault current flowing through the SCs under the same type of fault. The current limitation presents a slight difference among phases due to the difference in phase angle of each phase at the fault instant.

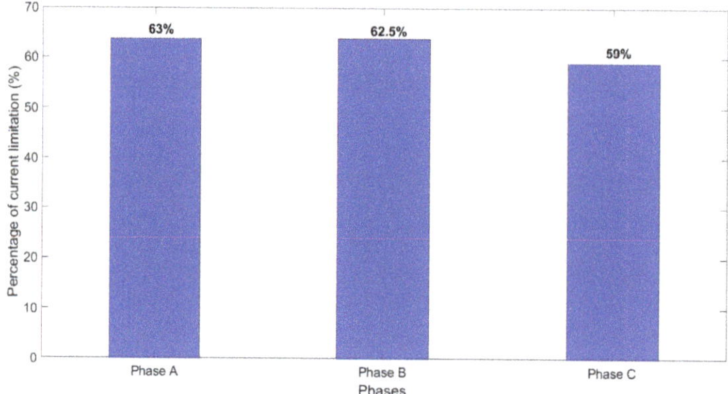

Figure 14. Current limitation percentage (%) of the SCs for phases A, B, and C compared to a conventional copper cable during a 3-Phase-to-ground fault at 50% of cable's length.

It is evident that the installation of SCs can lead to fault current reduction up to 62.5% of the prospective current flowing through a conventional copper cable, considering the same 3-Phase-to-ground fault.

3.3. Simulation Analysis of Fault Resistance Effect on the Quenching Process

In order to achieve the maximum benefit of the designed cable, its performance under a wide range of power system conditions should be comprehensively evaluated. In the available technical literature, several studies [45–50] have investigated the impact of the fault resistance R_f on the superconducting current limiters. However, there are no studies available assessing the impact of the fault resistance on the SCs and the fault current limitation that it provides. Therefore, in this Section the quenching process of the SCs is analyzed in accordance with the gradual increase in the fault resistance value. Simulation studies, which include 3-Phase-to-ground faults applied at the 50% of SCs length (considering different values of R_f), were conducted to study the relationship between R_f and the quenching process. Figures 15–20 show the corresponding waveforms of the quenching stage, the fault current signatures among the layers, the resistance and the temperature of the cable for $R_{f1} = 1\ \Omega$, $R_{f2} = 5\ \Omega$ and $R_{f3} = 10\ \Omega$, respectively.

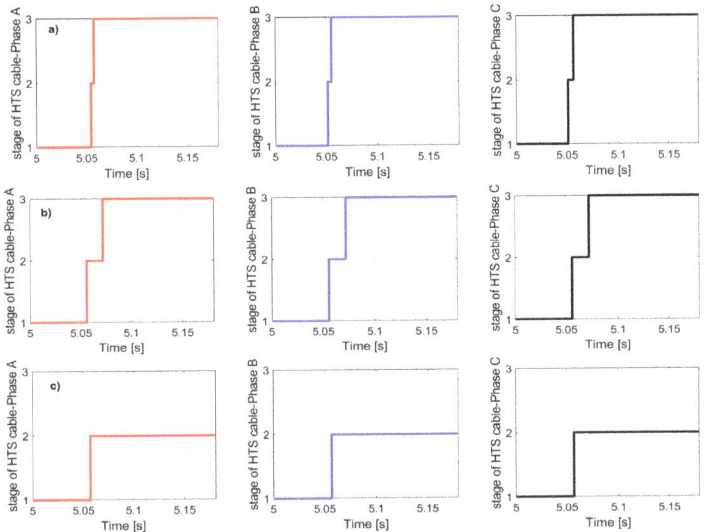

Figure 15. Stages of quenching process for phase A, B, and C for 3-Phase-to-ground fault at 50% of cable's length with (**a**) $R_{f1} = 1\ \Omega$, (**b**) $R_{f2} = 5\ \Omega$, (**c**) $R_{f3} = 10\ \Omega$.

Based on the results depicted in Figure 16, when fault resistance is $R_{f1} = 1\ \Omega$, the HTS tapes quench at the first half fault cycle and enter normal state (stage 3), as it can be seen in Figure 15a. The resistance and the temperature reach their maximum values at $t = 5.065\ s$, as shown in Figures 19a and 20a, respectively. Therefore, the current starts to flow through the stabilizer layer at $t = 5.065\ s$, 5 ms after the fault occurs. For the case of $R_{f2} = 5\ \Omega$, HTS tapes quench after one fault cycle (at $t = 5.082\ s$) and it is noticeable by Figure 15b that stage 2 lasts for a slightly longer period (few ms). Considering a fault resistance of $R_{f1} = 1\ \Omega$, SCs operate within stage 2 for 5.5 ms, while, for $R_{f2} = 5\ \Omega$, stage 2 lasts for 18 ms. Furthermore, for fault resistance $R_{f2} = 5\ \Omega$, the first fault current peaks depicted in Figure 18c,e,g are lower compared to the fault current peaks extracted during the fault with $R_{f1} = 1\ \Omega$, as a larger value of fault resistance results in lower fault currents. Regarding the case of $R_{f3} = 10\ \Omega$, the HTS tapes of the faulted phases quench, reaching only stage 2, without entering into normal state. Therefore, the maximum value of the SCs equivalent resistance is low, $R_{eq} = 0.058\ \Omega$ and the fault currents flow through the HTS layers. This behaviour indicates that the increase in the fault resistance value affects the quenching degree and consequently the current sharing between the HTS and stabilizer layers. Particularly, as it has already been analyzed (also reported in [28]), temperature increase plays a key role in the resulting value of the equivalent resistance and the quenching degree, which in turn is determined by the generated resistive heat, the magnitude, and the duration of the fault current. By observing, Figure 20a, it is obvious that for a 3-Phase-to-ground fault with $R_{f1} = 1\ \Omega$, the temperature of the superconductor exceeds the critical value T_C, reaching the maximum value of 250 K. When a 3-Phase-to-ground fault with $R_{f2} = 5\ \Omega$ occurs, the temperature exceeds the critical value $T_C = 92\ K$, but it is noticeable from Figure 20b, that the temperature reaches the maximum value of 250 K with a delay, which affects the quenching process. In the last case of $R_{f3} = 10\ \Omega$, the boundary condition $I_{fault} > I_C$ of quenching has been met. Although the temperature does not reach the critical value T_C, resulting in "incomplete quenching". The resistance of the HTS tapes reach low values, affecting the value of the equivalent resistance and the current distribution among the layers and resulting in small percentage of fault current limitation. The fault current flows mainly through the HTS layers.

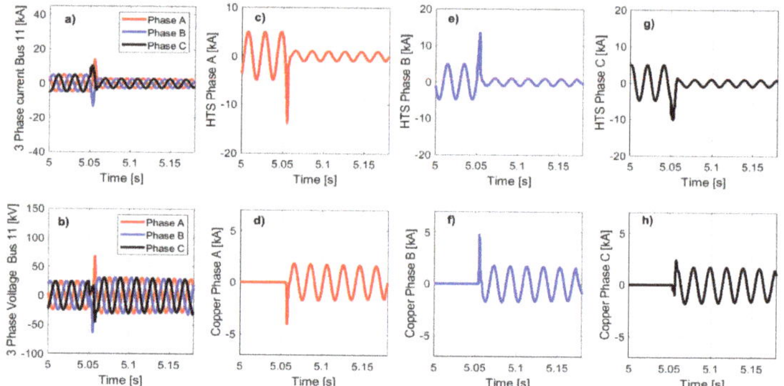

Figure 16. Fault current signatures for 3-Phase-to-ground fault at 50% of cable's length: with $R_{f1} = 1\,\Omega$ (**a**) phase currents at Bus 11, (**b**) phase voltages at Bus 11, (**c**) current in HTS layer of phase A, (**d**) current in copper layer of phase A, (**e**) current in HTS layer of phase B, (**f**) current in copper layer of phase B, (**g**) current in HTS layer of phase C, (**h**) current in copper slayer of phase C.

The results revealed that the fault resistance has a considerable impact on the SCs performance for the same type of fault, considering the same fault location. For instance, further increase in the fault resistance can lead to much lower fault currents, even below the critical current I_C, preventing SCs from quenching. This has been confirmed by Figure 18, where, during a 3-Phase-to-ground fault with $R_{f3} = 10\,\Omega$, the first peak of the fault current is below the critical current I_C, and therefore there is no quenching or fault current sharing between the two layers. Consequently, low values of fault resistance result in higher fault current (with respect to the critical current I_C), which lead to SCs quenching during the first half cycle, and therefore to greater fault current limitation capability (Figure 16). High fault resistance affects the quenching degree and jeopardizes the fault current limiting capability of the cable. This can be explained by the reduced allocation of fault current within different layers during current limitation mode, as fault current is predominately limited by the fault resistance value.

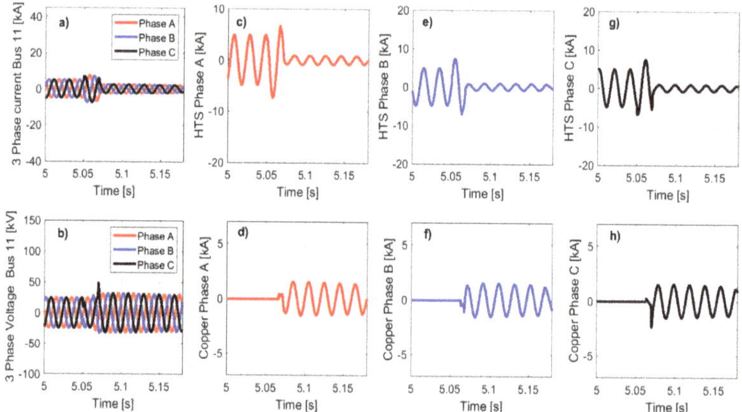

Figure 17. Fault current signatures for 3-Phase-to-ground fault at 50% of cable's length: with $R_{f2} = 5\,\Omega$ (**a**) phase currents at Bus 11, (**b**) phase voltages at Bus 11, (**c**) current in HTS layer of phase A, (**d**) current in copper layer of phase A, (**e**) current in HTS layer of phase B, (**f**) current in copper layer of phase B, (**g**) current in HTS layer of phase C, (**h**) current in copper slayer of phase C.

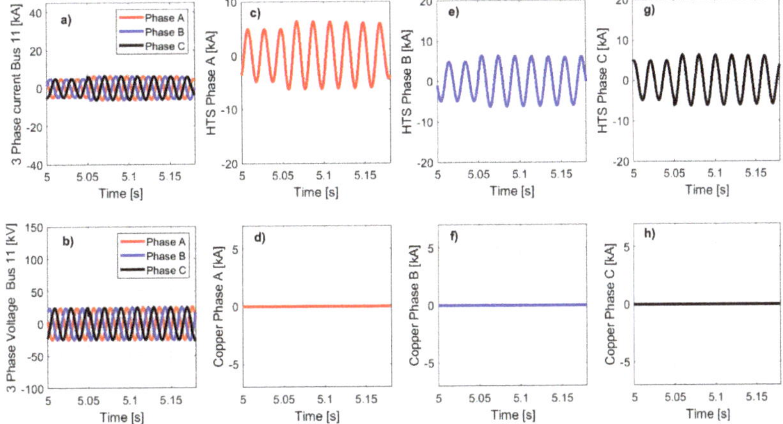

Figure 18. Fault current signatures for 3-Phase-to-ground fault at 50% of cable's length: with $R_{f3} = 10\,\Omega$ (**a**) phase currents at Bus 11, (**b**) phase voltages at Bus 11, (**c**) current in HTS layer of phase A, (**d**) current in copper layer of phase A, (**e**) current in HTS layer of phase B, (**f**) current in copper layer of phase B, (**g**) current in HTS layer of phase C, (**h**) current in copper slayer of phase C.

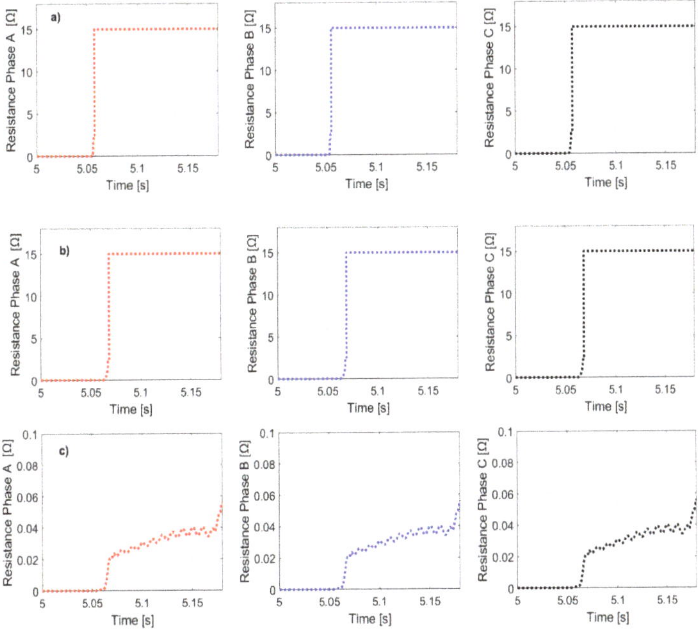

Figure 19. Equivalent resistance for phases A, B, and C for 3-Phase-to-ground fault at 50% of cable's length, with: (**a**) $R_{f1} = 1\,\Omega$, (**b**) $R_{f2} = 5\,\Omega$, (**c**) $R_{f3} = 10\,\Omega$.

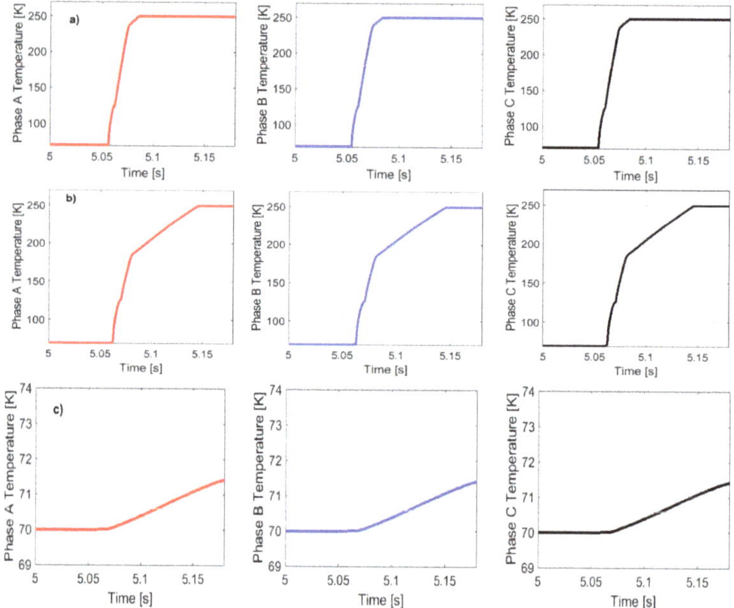

Figure 20. Temperature for phases A, B, and C for 3-Phase-to-ground fault at 50% of cable's length, with: (**a**) $R_{f1} = 1\ \Omega$ (**b**) $R_{f2} = 5\ \Omega$ (**c**) $R_{f3} = 10\ \Omega$.

4. Conclusions and Verifications

The comprehensive fault current characterization presented in this paper used a simplified, validated SCs model and highlighted the following key outcomes:

The operation of the SCs can be divided into three different stages: (i) the superconducting stage during the steady state operation of a power system, at which the SCs presents approximately zero resistance; (ii) the quenching process, which includes the partial resistive flux flow stage, reached when the fault current exceeds the critical current I_C and while temperature remains below the critical value T_C; and (iii) the highly resistive normal state which is reached once the temperature exceeds the critical value T_C. Furthermore, it has been found that during fault, the stabilizer layer can be used as a parallel path for the transient current, reducing heat generation, temperature rise, and protecting the cable from being damaged. To better investigate the feasibility of the copper stabilizer layer, the change of the copper layer resistivity has been modelled during the normal resistive mode, while the resistance of the HTS layer considered to be constant (i.e., set to its maximum value).

The performance of the SCs in limiting the fault currents was assessed through a number of fault scenarios. Simulation results revealed the impact of SCs on fault current magnitudes, under different type of faults, and as a consequent on the short-circuit level of the power systems. Specifically, it has been observed that within the first electric cycle, the magnitude of the fault current has been reduced from 15 kA to approximately 1.8 kA. Therefore, the installation of SCs introduces a challenge for the existing protection schemes due to their variable resistance, which leads to lower fault currents and higher voltage magnitude during transient conditions. In order to obtain a deeper insight of the fault current limiting capability of SCs, a comparative analysis has been conducted between the SCs and a conventional copper cable under the same fault conditions. The analysis revealed that for 3-Phase-to-g faults, the SCs model offers fault current limitation in the range of 60%, with respect to the prospective values. Therefore, the deployment of SCs increases the transmission efficiency due to the low resistance during the steady state and suppresses fault currents. The obtained results are well-aligned with relevant conducted studies such as [19,36,40].

The performance of SCs during transient conditions is determined by certain power system characteristics such as the prospective fault currents and fault resistance. In particular, simulation results showed that the increase in the fault resistance value impacts on the feasibility of the SCs, as it affects the quenching degree. It was revealed that the higher the fault resistance, the lower the prospective current and the percentage of current limitation. This was confirmed through the case of fault resistance equal to $R_{f3} = 10\ \Omega$, where the fault current is predominately limited by the fault resistance, affecting the increase in the resistivity of HTS tapes (which do not enter normal state) and the quenching process.

5. Future Work

After the fault current characterization and evaluation of SCs' performance during fault conditions, it has been identified that that there are many challenges that the protection schemes must take into consideration. As the simulation results revealed, the variable resistance of SCs, the reduced fault currents, the higher voltage magnitude during quenching stages, and the impact of high fault resistance values which jeopardize the quenching process, are factors which are anticipated to introduce challenges to the fault detection and classification methods, and by extension to the applicability of conventional protection schemes (e.g., over-current, distance protection, etc.) towards protecting SCs feeders. Considering the protection of future power grids (integrating SCs and inverter-connected generation), more research shall be steered towards the development of novel protection solutions which capture the particularities and distinctive features of SCs. For example, potential merits can arise from the utilization of learning-based methods for fault diagnosis on SCs, such as Deep-Learning techniques, which take advantage of the sequential relationships of the data and are able to handle long-term dependencies and correlated features that are important for fault diagnosis.

Author Contributions: Conceptualization and Investigation, E.T., A.E. and D.T.; Methodology E.T., D.T., A.E., A.D. and M.E.; Resources E.T., A.E., Q.H., W.Y. and M.E.; Software E.T., A.E. and D.T.; Analyses and interpretation of data E.T., A.E., D.T., W.Y. and A.D.; Supervision D.T., A.D., C.B. and Q.H.; Visualization E.T.; Writing-original draft E.T.; Writing-review and editing D.T., A.E., A.D., C.B., Q.H., M.E. and W.Y. All authors have read and agreed to the published version of the manuscript.

Funding: This research received no external funding.

Conflicts of Interest: The authors declare no conflict of interest.

References

1. Zhang, Z.; Venuturumilli, S.; Zhang, M.; Yuan, W. Superconducting cables—Network feasibility study work package 1. In *Innovation, Next Generation Networks*; Western Power Distribution: Castle Donington, UK, 2016; Volume 9, pp. 8–15.
2. Blair, S.M.; Booth, C.D.; Burt, G.M. Current-Time Characteristics of Resistive Superconducting Fault Current Limiters. *IEEE Trans. Appl. Supercond.* **2012**, *22*, 5600205. [CrossRef]
3. Blair, S.M.; Booth, C.D.; Elders, I.M.; Singh, K.N.; Burt, G.M.; McCarthy, J. Suprconducting fault current limiter application in power dense marine electrical system. *IET Elect. Sust. Transp.* **2011**, *1*, 93–102. [CrossRef]
4. Blair, S.M.; Singh, N.K.; Booth, C.D. Operational control and protection implications of fault current limitation in distribution networks. In Proceedings of the 2009 44th International Universities Power Engineering Conference (UPEC), Glasgow, UK, 1–4 September 2009; pp. 1–5.
5. Albany Power Cables. Available online: https://www.energy.gov/sites/prod/files/oeprod/DocumentsandMedia/Albany__03_05_08.pdf (accessed on 6 June 2020).
6. Kojima, H.; Kotari, M.; Kito, T.; Hayakawa, N.; Hanai, M.; Okubo, H. Current Limiting and Recovery Characteristics of 2 MVA Class Superconducting Fault Current Limiting Transformer (SFCLT). *IEEE Trans. Appl. Supercond.* **2010**, *21*, 1401–1404. [CrossRef]

7.   Liang, F.; Yuan, W.; Baldan, C.A.; Zhang, M.; Lamas, J.S. Modeling and Experiment of the Current Limiting Performance of a Resistive Superconducting Fault Current Limiter in the Experimental System. *J. Supercond. Nov. Magn.* **2015**, *28*, 2669–2681. [CrossRef]

8.   Stuart, W.; Strickland, N. Critical Current Characterization of THEVA Production 2G HTS Superconducting Wire. Available online: https://figshare.com/articles/dataset/Critical_current_characterisation_of_THEVA_pre-oduction_2G_HTS_superconducting_wire/375932 (accessed on 7 June 2020).

9.   Mark, F.M. Ampacity Project-Worldwide First Superconducting Cable and Fault Current Limiter intallation in a German City center. In Proceedings of the 22nd International Conference on Electricity Distribution, Stockholm, Sweden, 10–13 June 2013.

10. Shibata, T.; Watanabè, M.; Suzawa, C.; Isojima, S.; Fujikami, J.; Sato, K.; Ishii, H.; Honjo, S.; Iwata, Y. Development of high temperature superconducting power cable prototype system. *IEEE Trans. Power Deliv.* **1999**, *14*, 182–187. [CrossRef]

11. Sohn, S.; Lim, J.; Yang, B.; Lee, S.; Jang, H.; Kim, Y.; Yang, H.; Kim, D.; Kim, H.; Yim, S.; et al. Design and development of 500m long HTS cable system in the KEPCO power grid, Korea. *Phys. C Supercond.* **2010**, *470*, 1567–1571. [CrossRef]

12. Ballarino, A.; Bruzek, C.E.; Dittmar, N.; Giannelli, S.; Goldacker, W.; Grasso, G.; Grilli, F.; Haberstroh, C.; Hole, S.; Lesur, F.; et al. The BEST PATHS Project on MgB2Superconducting Cables for Very High Power Transmission. *IEEE Trans. Appl. Supercond.* **2016**, *26*, 1–6. [CrossRef]

13. Kalsi, S.S. HTS Superconductors. *Appl. High Temperature Supercond. Electr. Power Equip.* **2011**, *9*, 219–259.

14. Muncho, J.M. Development trend of Superconducting Cable Technology for Power Transmission Applications. *World Congr. Eng. Comput.* **2015**, *1*, 1–4.

15. Snitcher, G.; Kalsi, S.S.; Manlief, M.; Schwall, E.E.; Sidi-Yekhief, A.; Ige, S.; Francavilla, T.L.; Gusber, D.U. High-filed warm-bore HTS conduction cooled magnet. *IEEE Trans. Appl. Supercond.* **1999**, *9*, 553–558. [CrossRef]

16. Luiz, A. *Applications of High-Tc Superconductivity*; IntechOpen: Rijeka, Croatia, 2011; Volume 3, pp. 46–72.

17. Akhtar, M.J. 100 Years of Superconductivity (1911–2011). *Nuclues* **2011**, *48*, 261.

18. Huebener, R.R. Conductors, Semiconductors, Superconductors: An introduction tosolid state physics. *Univ. Tub.* **2004**, *3*, 120–141.

19. Young, M.A. An investigation of the current distribution in the trixial cable and its operational impacts on power systems. *IEEE Trans. Appl. Supercond.* **2005**, *15*, 1751–1754. [CrossRef]

20. Bang, J. Critical current, criticial temperature anf magnetic filed based EMTDC model components for HTS power cables. *IEEE Trnas. Appl. Supercond.* **2007**, *17*, 1726–1729. [CrossRef]

21. Yoon, D.-H. A Feasibility Study on HTS Cable for the Grid Integration of Renewable Energy. *Phys. Procedia* **2013**, *45*, 281–284. [CrossRef]

22. Malozemoff, A.P. High Temperature Seperconducting HTS AC Cables for Power Grid Applications. In *Superconductors in Power Grid*; Woodhead Publishing: Sawston, UK, 2015; pp. 133–188.

23. Jipping, J.; Mansoldo, A.; Wakefield, C. The impact of HTS cables on power flow distribution and short-circuit currents within a meshed network. In Proceedings of the 2001 IEEE/PES Transmission and Distribution Conference and Exposition, Developing New Perspectives (Cat. No.01CH37294), Atlanta, GA, USA, 2 November 2002; pp. 736–741. [CrossRef]

24. Maruyama, O.; Nakano, T.; Mimura, T.; Morimura, T.; Masuda, T.; Takagi, T.; Yagi, M. Fundamental Study of Ground Fault Accident in HTS Cable. *IEEE Trans. Appl. Supercond.* **2017**, *27*, 1–5. [CrossRef]

25. Ishiyama, A.; Wang, X.; Ueda, H.; Uryu, T.; Yagi, M.; Fujiwara, N. Over-Current Characteristics of 275-kV Class YBCO Power Cable. *IEEE Trans. Appl. Supercond.* **2010**, *21*, 1017–1020. [CrossRef]

26. Maguire, J.; Folts, D.; Yuan, J.; Henderson, N.; Lindsay, D.; Knoll, D.; Rey, C.; Duckworth, R.; Gouge, M.; Wolff, Z.; et al. Status and progress of a fault current limiting hts cable to be installed in the con edison grid. *AIP Conf. Proc.* **2010**, *1218*, 445.

27. Maguire, J.; Folts, D.; Yuan, J.; Lindsay, D.; Knoll, D.; Bratt, S.; Wolff, Z.; Kurtz, S. Development and Demonstration of a Fault Current Limiting HTS Cable to be Installed in the Con Edison Grid. *IEEE Trans. Appl. Supercond.* **2009**, *19*, 1740–1743. [CrossRef]

28. Liang, S.; Ren, L.; Ma, T. Study on quenching characteristics and resistance equievalent estimation method of second-generation High Temperature Superconducting tape under different overcurrent. *Materials* **2019**, *12*, 2374. [CrossRef]

29. Doukas, D.I.; Chrysochos, A.I.; Labridis, D.P.; Harnefors, L.; Velotto, G. Coupled electro-mechanical transient analysis of superconducting DC transmission system using FDTD and VEM modelling. *IEEE Trans. Appl. Supercond.* **2017**, *27*, 1–8. [CrossRef]

30. Fukui, S.; Ogawa, J.; Suzuki, N.; Oka, T.; Sato, T.; Tsukamoto, O.; Takao, T. Numerical Analysis of AC Loss Characteristics of Multi-Layer HTS Cable Assembled by Coated Conductors. *IEEE Trans. Appl. Supercond.* **2009**, *19*, 1714–1717. [CrossRef]

31. Yao, M. Numerical Modelling of Superconducting Power Cables with Second Generation High Temperature Superconductors. Ph.D. Thesis, University of Edinburgh, Scotland, UK, 2019.

32. Su, R.; Shi, J.; Yan, S.; Li, P.; Wng, W.; Hu, Z.; Zhang, B.; Tnag, Y.; Ren, L. Numerical model of HTS cable and its electri-thermal properties. *IEEE Trans. Appl. Supercond.* **2019**, *29*, 1–5.

33. Shaghai Superconductors. Available online: http://www.shsctec.com/en/introduce/89 (accessed on 10 June 2020).

34. Polak, M.; Krempaský, L.; Chromik, Š.; Wehler, D.; Moenter, B. Magnetic field in the vicinity of YBCO thin film strip and strip with filamentary structure. *Phys. C Supercond.* **2002**, *372*, 1830–1834. [CrossRef]

35. Du, H.-I.; Kim, T.-M.; Han, B.-S.; Hong, G.-H. Study on Verification for the Realizing Possibility of the Fault-Current-Limiting-Type HTS Cable Using Resistance Relation With Cable Former and Superconducting Wire. *IEEE Trans. Appl. Supercond.* **2015**, *25*, 1–5. [CrossRef]

36. Nguyen, T.-T.; Lee, W.-G.; Lee, S.-J.; Park, M.; Kim, H.-M.; Won, D.; Yoo, J.; Yang, H.S. A Simplified Model of Coaxial, Multilayer High-Temperature Superconducting Power Cables with Cu Formers for Transient Studies. *Energies* **2019**, *12*, 1514. [CrossRef]

37. Choi, Y.; Kim, J.; Kim, Y.; Ko, T.; Lee, H.; Kim, J.; Choi, Y. Quench and recovery characteristics of second-generation high-temperature superconducting GdBCO coated conductor with various patterns of stabilizers. *Results Phys.* **2018**, *10*, 222–226. [CrossRef]

38. Fu, Y.; Tsukamoto, O.; Furuse, M. Copper stabilization of YBCO coated conductor for quench protectio. *IEEE Trans. Appl. Supercond.* **2003**, *13*, 1780–1783. [CrossRef]

39. Zhu, J.; Zhang, Z.; Zhang, M.; Zhang, M.; Qiu, M.; Yuan, W. Electric Measurement of the Critical Current, AC Loss, and Current Distribution of a Prototype HTS Cable. *IEEE Trans. Appl. Supercond.* **2013**, *24*, 1–4. [CrossRef]

40. Elshiekh, M.; Zhang, M.; Ravindra, H.; Chen, X.; Venuturumilli, S.; Huang, X.; Schoder, K.; Steurer, M.; Yuan, W. Effectiveness of Superconducting Fault Current Limiting Transformers in Power Systems. *IEEE Trans. Appl. Supercond.* **2018**, *28*, 1–7. [CrossRef]

41. Torii, S.; Kasahara, H.; Akita, S. Temperature dependence of critical current density of AC superconductor and the effect on AC quench current degradation. *IEEE Trans. Appl. Supercond.* **1995**, *5*, 369–372. [CrossRef]

42. Hong, Z.; Campbell, A.M.A.; Coombs, T. Numerical solution of critical state in superconductivity by finite element software. *Supercond. Sci. Technol.* **2006**, *19*, 1246–1252. [CrossRef]

43. Serway, R.A.; Kirkpatrick, L.D. Physics for Scientists and Engineers with Modern Physics. *Phys. Teach.* **1988**, *26*, 254–255. [CrossRef]

44. Jin, T.; Hong, J.-P.; Zheng, H.; Tang, K.; Gan, Z.-H. Measurement of boiling heat transfer coefficient in liquid nitrogen bath by inverse heat conduction method. *J. Zhejiang Univ. A* **2009**, *10*, 691–696. [CrossRef]

45. Colombus HTS Cables. Available online: https://www.energy.gov/sites/prod/files/oeprod/DocumentsandMedia/columbus_03.05.08.pdf (accessed on 7 June 2020).

46. Barzegar-Bafrooei, M.R.; Foroud, A.A. Impact of fault resistance and fault distance on fault current reduction ratio of hybrid SFCL. *Int. Trans. Electr. Energy Syst.* **2017**, *27*, e2409. [CrossRef]

47. Mansour, D.-E.A. Effect of fault resistance on the behavior of superconducting fault current limiter in power systems. In Proceedings of the 2014 IEEE International Conference on Power and Energy (PECon), Kuching, Malaysia, 1–3 December 2014; pp. 212–216. [CrossRef]

48. Kalsi, S.S. *Applications of High Temperature Superconductors to Electric Power Equipment*; John Wiley & Sons: Hoboken, NJ, USA, 2011; pp. 173–217. [CrossRef]

damage or even system blackouts [1–7], which has greatly harmed the safe operation of the power system. Therefore, it is urgent to enhance the safety defense ability of the power generation-side protection system.

In order to guarantee the protection performance and achieve the effective coordination of the generation-side protection with the system operation requirements, the current research mainly focuses on the principle improvement of the generator protection related to the network and its coordination with the automatic control system. The literature [8,9] proposes an improved out-of-step protection method based on equal-area-criterion theory, which can more accurately discriminate the transient stability state and the out-of-step fault state. The literature [10] proposes a fault-current-based stator ground fault protection method which constructs the protection criterion and determines the generator tripping strategy according to the fault degree. In order to improve the regulation and control ability of the generator excitation regulation system and the automatic control system, the study in [11] standardizes the coordination relationship of the loss-of-excitation protection and the low excitation limitation. In addition, the literature [12] gives a coordination scheme of the over-speed protection and the high-frequency tripping measures of the automatic control system. All the above methods can improve the power support ability of the generation-side under fault conditions. However, due to lack of more abundant local fault information and appropriate information interaction between the generators and between the generation and grid sides, the protection performance cannot be further improved.

With the rapid development of communication technology and the application of the wide area measurement system [13–15], the hierarchical transmission network protection system has been widely concerned, studied and applied in some engineering. From this, a variety of new protection technologies have been developed. Among them, wide area current differential protection [16,17] can achieve the effective backup protection coordination under system disturbance situations. Wide area back-up protection [18–20] can quickly accomplish the fault component identification, tripping decision and other functions through information interaction by the wide area communication network. Station area integrated protection [21,22] acquires the multi-information on the station layer at the same time, and the protection function coordination can be realized on the basis of completing each independent protection function. However, the multi-information fusion-based hierarchical protection technology is still lacking in the generation-side protection system, which restricts the safe operation of the main pieces of equipment in the power plants and their coordination with the power system under the modern complex power grid environment.

In order to overcome the protection performance limitations and the multi-generator protection coordination problems, on the basis of the existing local protection system, condition monitoring system and power control system, the hierarchical protection system is established through multi-information interaction and fusion among the above systems, the generation units and the power grid. This paper discusses the structure and function of the hierarchical power generation-side protection system. The local layer multi-information fusion-based comprehensive protection method and the multi-generators information fusion-based hierarchical protection method are taken as examples to illustrate the protection principle construction method and the actual application mode. Finally, the key issues of how to complete the hierarchical power generation-side protection system are discussed. Compared with the current protection methods, the multi-information fusion-based hierarchical protection methods can successfully improve the protection sensitivity and can adaptively determine the protection action characteristic and tripping strategy according to the system operating conditions and other generators' fault conditions. They can effectively enhance the protection performance and provide reliable guarantee for the safe operation coordination of the generation and grid sides.

2. Structure and Function of the Hierarchical Power Generation-Side Protection System

The goal of the multi-information fusion-based hierarchical power generation-side protection system is to enhance the protection performance and satisfy the security and co-ordinated operation requirements between the generation and grid sides. The hierarchical protection system combines the information from the condition monitoring system, power generation control system, local protection system and power grid dispatch and control system to accomplish the information interaction and fusion. On this basis, the relatively independent design and function of the current generation-side protection system can be changed.

Figure 1 shows the structure diagram of the multi-information fusion-based hierarchical power generation-side protection system. It is composed of a local layer, station layer and system layer.

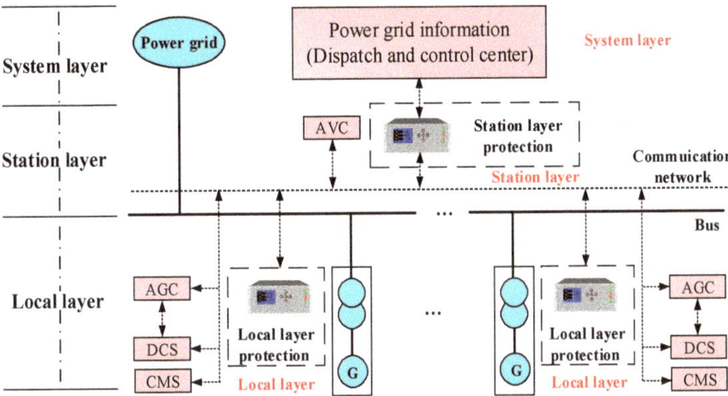

Figure 1. Structure diagram of the multi-information fusion-based hierarchical power generation-side protection system.

As for the local layer, it is composed of the local generation units' protection devices to achieve the local protection function. However, its protection information source is not limited to the local measurement data, it also includes the fused information from other generation units' protection, automatic voltage control system (AVC), automatic generation control system (AGC), distributed control system (DCS) and condition monitoring system (CMS) through the communication network supported by the station layer. In the hierarchical protection system, the local layer protection can not only satisfy the rapidity and reliability requirements of the local protection function, but also use the condition information and control information to master the healthy condition of the generation units and improve the protection performance. In addition, the local layer protection undertakes the safety protection instructions assigned by the station layer, such as warning, tripping the generator, adjusting the generation output, etc.

The station layer is a new layer in the generation-side protection system on the basis of the existing protection devices. It can provide a transmission platform for information interaction and fusion between the generation units and the power grid, serving as the information management center of the whole protection system. It can satisfy the local layer protection information requirements and realize the station layer protection function. More importantly, it can enhance the performance of the grid-related generation-side protection, and better adapt to the cooperative and safe operation between the generation and grid sides. In addition, the station layer also accomplishes the unified voltage control and power output dispatching of the generators in the power plant according to the instructions from the system layer. The existing dispatch and control center adopts the "one to one" power dispatching mode, which is directly targeted at the specific generator. Due to lack of information interaction, the healthy condition of the generator cannot be mastered by the dispatch and control center. In the situation that the generator output is adjusted but is being made to enter an unhealthy operation condition, if the protection acts and the generator is quickly tripped, the system power shortage will become worse. However, if the protection does not act in time, the generator may be damaged. After introducing the station layer, the health condition constraints of the generators and the safety operation requirements of the power grid can be fully reflected. Thus, the dispatching and controlling strategy can effectively meet the real-time safety demand of the whole system.

The system layer is composed of the power grid dispatch and control center. It can monitor and grasp the voltage level of each node in the system and the power supply demand of the users. Through integrating the information from the whole network, it assigns the voltage regulation, power dispatching and other instructions which can meet the requirements of the whole power grid to the station layer. It also summarizes and analyzes the operation condition and protection information of the power grid-side and sends them to the generation-side. Then the station layer can attain the goal of remote information interaction, unified control and collaborative protection between the generation and grid sides.

The multi-information fusion-based hierarchical power generation-side protection system does not change the basic configuration of the existing protection devices and automatic monitoring devices. It only needs to add the communication network and set up the station layer protection device to complete the basic structure of the protection system, which is easily accomplished. The hierarchical protection system can make full use of the multi-information fusion and protection layering, which can enrich the protection information source, satisfy the protection coordination with the system operating environment and guarantee the security defense ability.

3. Several Examples of the Hierarchical Generation-Side Protection System Construction Methods

The existing generation-side protection system is constructed with local protection information. Due to lack of information fusion with CMS, the protection information is not complete enough to fully master the real-time security condition of the corresponding generation units. Moreover, since the protection information is not abundant enough, the protection performance cannot be further improved. In addition, since information interaction between the generation units and between the generation and grid sides cannot be achieved, the existing protection methods and action strategies are unable to attain the coordination among the protection units under the fault forms which affect multi-generators. In this section, the multi-information fusion-based comprehensive local layer protection construction method and the multi-generators information fusion-based hierarchical protection construction method are proposed as examples to illustrate the hierarchical protection construction method and the application mode.

3.1. Multi-Information Fusion Based Comprehensive Local Layer Protection

In order to solve the problem that the existing local information-based generation-side protection system cannot fully master the real-time security condition and cannot further enhance the protection performance, the multi-information fusion-based hierarchical protection system is used to improve the local layer protection principles. The relevant information such as electrical, mechanical and condition monitoring information which can reflect the safety condition of the generation units is introduced to enhance the protection sensitivity. In this section, the comprehensive single transverse differential protection criterion is taken as an example to illustrate the multi-information fusion-based comprehensive local layer protection construction method.

3.1.1. Review of the Existing Single Transverse Differential Protection

For large generators, their internal structures are complex and their operation conditions are changeable. Thus, the internal fault types are various and the fault forms are complex. For the slight turn-to-turn faults, due to the short-circuit current not being large, the protection sensitivity is usually low [23]. However, if the protection fixed value is blindly lowered to improve the protection sensitivity, the unbalanced current under external faults may lead to protection maloperation, causing unnecessary generator tripping and endangering the safe and stable operation of the power system.

The existing single transverse differential protection for the large generators usually uses the phase current braking criterion [24]. The braking criterion can prevent protection maloperation. However, the existing research [25,26] shows that the changing rules are quite different between the phase current and the transverse differential unbalanced current, and they cannot establish a direct relationship. The protection method cannot avoid the unbalanced current effectively under different operation modes, which can easily cause protection maloperation. In addition, since the phase current braking criterion is used, the braking current is relatively large under internal faults. Thus, the protection sensitivity is reduced. Therefore, it is necessary to integrate other protection information which can fully reflect the transverse differential unbalanced current to improve the existing protection braking criterion.

3.1.2. Air Gap Electromotive Force Braking Criterion for Single Transverse Differential Protection

In order to solve the problem that the widely used phase current braking criterion cannot effectively reflect the transverse differential unbalanced current, the generation mechanism of the transverse differential unbalanced current is analyzed to find the exact electric quantity which can fully reflect it. Then a novel protection criterion is constructed.

Due to various reasons such as design, manufacturing, installation and operation, the internal structure of a generator cannot be absolutely symmetrical, which makes the inductance parameters (including self-inductance and mutual inductance) of each branch winding different. Thus, the imbalance of the main air gap flux is caused, and the imbalance of the stator winding electromotive force is caused. On this basis, the unbalanced current is generated on the connecting line of the neutral points (O_1 and O_2); its value is related to the internal structure asymmetry degree and the main air gap flux strength. The schematic of the multi-branches generator stator windings is shown in Figure 2.

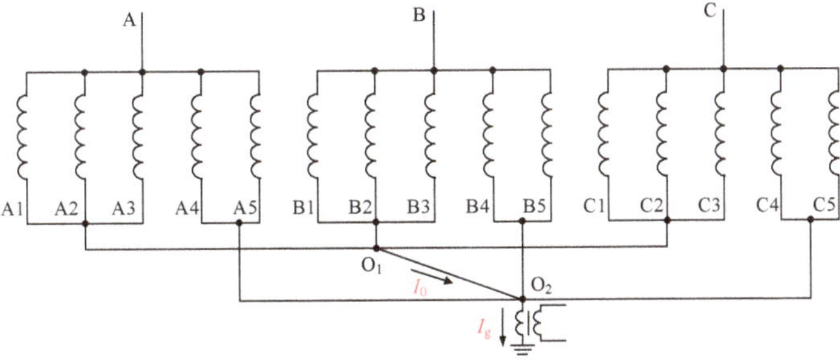

Figure 2. Schematic of the multi-branch generator stator windings.

Large generators usually use the small current grounding mode, and the low-voltage windings of the step-up transformer are usually connected by the delta type. Thus, the value of the ground current flowing into the earth from the neutral point I_g is very small. Compared with the value of the transverse differential unbalanced current I_0, it can be ignored. Based on this, in Figure 2, the parallel branches in the same phase which are connected at the same neutral point are equivalent to a single branch, and the equivalent circuit diagram is shown in Figure 3.

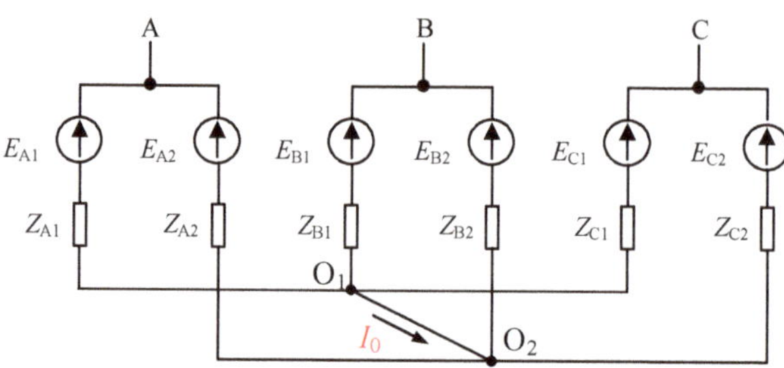

Figure 3. Equivalent circuit of the generator stator windings.

In Figure 3, E_{A1}, E_{A2}, E_{B1}, E_{B2}, E_{C1}, E_{C2} represent the comprehensive equivalent electromotive force of each branch and Z_{A1}, Z_{A2}, Z_{B1}, Z_{B2}, Z_{C1}, Z_{C2} represent the comprehensive equivalent internal impedance of each branch. Since the transverse differential unbalanced current I_0 has the same property as the zero-sequence current, the three-phase branches connected at the same neutral point can be further combined into an equivalent branch, as shown in Figure 4.

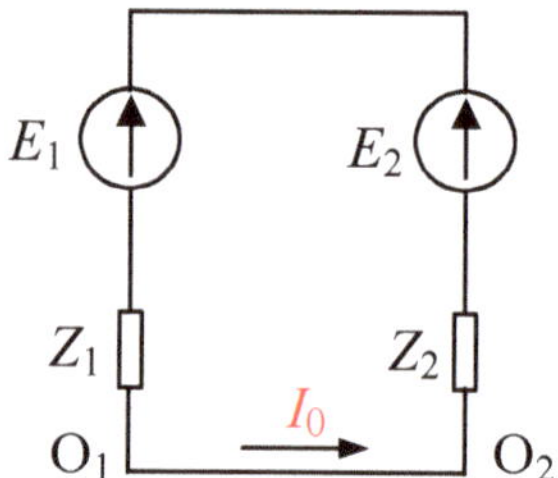

Figure 4. Simplified equivalent circuit of the generator stator windings.

In Figure 4, E_1 and E_2 are the equivalent induced electromotive forces, reflecting the asymmetric degree of the three-phase induced electromotive force. Z_1 and Z_2 are the equivalent internal impedances. When the equivalent induced electromotive forces are not equal, the transverse differential unbalanced current on the neutral point line is produced as

$$I_0 = \frac{|E_1 - E_2|}{Z_1 + Z_2} = \frac{\Delta E}{Z_1 + Z_2} \tag{1}$$

where ΔE is the unbalanced induction electromotive force of the stator windings. Since it is generated by the alternations of the main air-gap flux with time, then

$$\Delta E = 4.44 f N k_{N1} \Delta \Phi_m \tag{2}$$

where f is the power frequency, N is the stator winding turns, k_{N1} is the winding coefficient, and $\Delta \Phi_m$ is the unbalanced main air gap flux of the stator windings which satisfies the following relationship

$$\Phi_m \propto LI \tag{3}$$

where L is the comprehensive equivalent inductance of the generator windings, and I is the comprehensive equivalent current of the generator stator and rotor. Thus, the relationship between the unbalanced main air gap flux $\Delta \Phi_m$ and the main air gap flux Φ_m can be expressed as

$$\Delta \Phi_m \propto \frac{\Delta L}{L} \Phi_m \tag{4}$$

where ΔL means the winding inductance parameter difference of every branch, it reflects the asymmetry degree of the generator internal structure.

Once a generator is manufactured, its internal structure asymmetry remains unchanged. Combined with Equations (1)–(4), it can be analyzed that the transverse differential unbalanced current I_0 is proportional to the main air-gap flux amplitude Φ_m. According to the electric machinery theory, the main air-gap flux amplitude Φ_m is proportional to the air gap electromotive force E_δ. Thus, E_δ can be introduced to construct the protection criterion to monitor the change of the main air-gap flux. In addition, the monitoring result can be taken as the assistant criterion of the single transverse differential protection. The corresponding protection criterion is

$$\begin{cases} I_0 > I_{0n} & E_\delta \leq E_{\delta n} \\ I_0 > \frac{E_\delta}{E_{\delta n}} I_{0n} & E_\delta > E_{\delta n} \end{cases} \tag{5}$$

In Equation (5), I_0 is the measured transverse differential current, E_δ is the measured air gap electromotive force; the fixed value of $E_{\delta n}$ is set as the lowest air gap electromotive force when the generator operates normally and the fixed value of I_{0n} is set as the corresponding unbalanced current under the lowest air gap electromotive force. Among them, E_δ cannot be directly measured, it can be calculated by the following formula

$$\dot{E}_\delta = \dot{U}_g + \dot{I}_g (R_a + jX_\sigma) \tag{6}$$

where R_a and X_σ are the leakage resistance and the leakage reactance of the generator stator windings, respectively, they are the inherent parameters of the generator. $\dot{U}_g$ and $\dot{I}_g$ are the terminal voltage and current of the generator respectively, which can be measured by potential transformers and current transformers.

3.1.3. Comprehensive Criterion for the Hierarchical Single Transverse Differential Protection

The existing phase current braking criterion of the single transverse differential protection can effectively prevent the protection maloperation when external fault occurs. However, at the same time, the protection sensitivity under internal faults is reduced. As for the air gap electromotive force braking criterion, when internal turn-to-turn faults occur, due to the values of the voltage and current at the generator terminal generally do not have obvious change, the air gap electromotive force will not change. However, since the transverse differential current increases, the protection can act correctly. Even in some internal fault cases that the current at the generator terminal increases obviously, since the air gap electromotive force is also related to the terminal voltage and its fluctuation amplitude is generally small, the amplitude of the air gap electromotive force can be limited to a certain extent, and it can be prevented from entering the braking area.

In order to improve the performance of the local layer protection, based on the hierarchical protection system, the multi-information fusion-based method is used to construct the comprehensive protection criterion by combining the proposed air gap electromotive force braking criterion with the existing phase current braking criterion. Furthermore, the logical solution of the comprehensive protection criterion is shown in Figure 5.

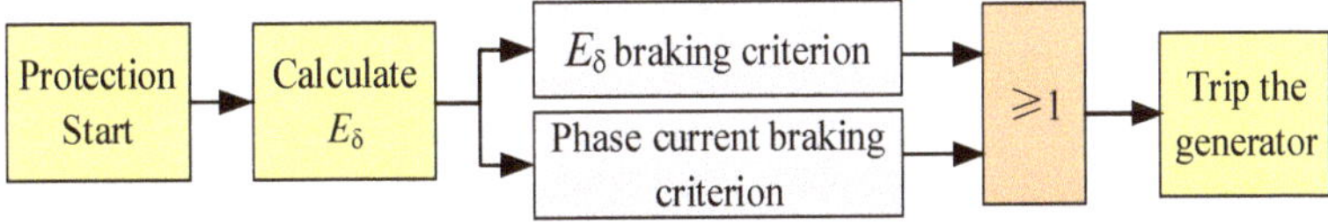

Figure 5. Logical solution of the hierarchical single transverse differential protection comprehensive criterion.

The two criterions adopt the "or door" mode. When any criterion satisfies the action situation, the comprehensive single transverse differential protection will act. The two criterions can play to their respective advantages. The air gap electromotive force braking criterion can enlarge the protection scope and enhance the protection sensitivity for the internal small turn-to-turn faults, and the phase current braking criterion can prevent the protection maloperation when an external fault occurs. Thus, the protection performance can be effectively improved.

3.1.4. Simulation Analysis of Protection Action Conditions under Generator Internal Faults

In order to verify the correctness of the above theoretical analysis and the effectiveness of the proposed protection scheme, the relevant simulation analysis is carried out to test the transverse differential protection action conditions under internal faults. As for the large generator internal faults, the fault forms are large in number, and the electromagnetic transient relationship is very complex. Since the generator models in commonly used power system simulation software (such as ATP, EMTP, EMTDC, MATLAB, etc.) are packaged, the transient simulation calculation for internal faults cannot be carried out. Thus, the relevant simulation analysis is carried out based on the "Large hydro-generator internal fault simulation, main protection analysis and design system" developed by our research team. The detailed derivation and introduction of the model used in this software are presented in the literature [27,28]. The software system has been successfully applied to the internal fault simulation of the Three-Gorges hydropower plant and other large hydro-generators. The experimental results and operating experience show that the simulation results can meet the engineering requirements.

To verify the viewpoint that the phase current cannot reflect the transverse differential unbalanced current, but the air gap electromotive force can, the 800 MW generator of a hydropower plant is taken as an example, and the relationships of the transverse differential unbalanced current with the phase current and the air gap electromotive force during normal operation and external short circuit faults are simulated. Set the following simulation conditions: rated voltage and rated power factor with 10%, 40%, 70%, 100% load; the rated voltage and rated current with the power factors are 0.8, 0.7 and 0.6, respectively; the rated current and rated power factor with the terminal voltage is 0.1, 0.3, 0.5, 0.7, 0.9 times the rated voltage; The three-phase short-circuit fault current at the generator terminal is 10%, 40%, 70% and 100% of the rated current, respectively. For each condition, the transverse differential current I_0, the terminal voltage U_g and the terminal phase current I_g are measured, and the air gap electromotive force E_δ is calculated. The relation diagrams of I_0 with E_δ and I_0 with I_g are shown in Figure 6.

The simulation results show that there is a linear relationship between I_0 and E_δ under both the normal operation conditions and the external fault conditions. However, there is no clear relationship between I_0 and I_g. The two are approximately linear only in the cases of three-phase short-circuit faults at the generator terminal corresponding to the data points in the circle of the Figure 6b. Therefore, it can be proved that the air gap electromotive force can reflect the transverse differential unbalanced current more effectively.

According to the winding structure of the case generator stator, the software generates the possible internal short-circuit fault set. Among them, "slot internal faults (SIF)" refers to the short circuit faults caused by insulation damage of the upper and lower coils in the same slot, and "terminal internal faults (TIF)" refers to the short circuit faults caused by insulation damage at the intersection of the coil terminals. According to the generator parameters and the simulation results, the parameters of the protection action curve can be calculated. The internal fault set and the corresponding protection action results are shown in Table 1.

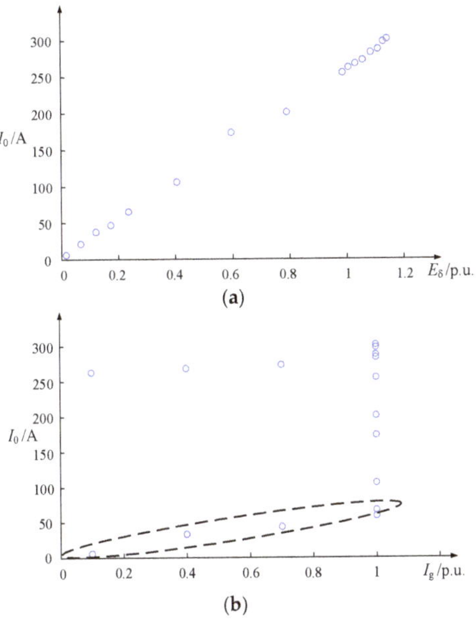

Figure 6. Simulation results of the generator transverse differential unbalanced current: (**a**) the relation diagram of the transverse differential unbalanced current with the air gap electromotive force; (**b**) the relation diagram of the transverse differential unbalanced current with the phase current.

Table 1. Statistical results of the generator internal fault set and the protection action condition results.

	Internal Fault Set				Comprehensive Protection Criterion				Existing Protection Criterion			
Fault Location	Same Branch	Different Branch	Different Phase	Total	Same Branch	Different Branch	Different Phase	Total	Same Branch	Different Branch	Different Phase	Total
SIF	432	48	360	840	428	48	360	836	420	48	360	828
TIF	768	3792	11,400	15,960	765	3768	11,376	15,909	761	3723	11,342	15,826
Total	1200	3840	11,760	16,800	1193	3816	11,736	16,745	1181	3771	11,702	16,654

In order to compare the protection performance of the existing phase current braking criterion and the proposed comprehensive protection criterion, the action results of the two protection criterions are listed in Table 1. By analyzing the statistical results, it can be seen that there are 16,800 forms of possible internal faults for the case generator. For the existing phase current braking protection criterion, the protection can act correctly under 16,654 forms of fault conditions. However, the comprehensive protection criterion can correctly act under 16,745 forms of fault conditions, which can supplement 91 kinds of faults compared with the existing protection criterion. These fault events are the slight internal faults for which the fault features are not obvious. However, since the large hydropower generator has a complex internal structure, these kind of fault forms have higher occurrence probability. If not removed in time, their long-term existence will also burn out the stator insulation and cause more serious faults. By introducing the air gap electromotive force braking criterion, the multi-information fusion-based comprehensive criterion of the single transverse differential protection can effectively reduce the protection action setting value under internal faults and the protection sensitivity is enhanced with the same braking ability under external faults. Thus, the protection performance is successfully improved.

3.2. Multi-Generators Information Fusion Based Hierarchical Protectio

The existing generation-side protection system only uses the local protection information, and the information interaction between generation units and interaction between source and grid cannot be achieved. However, as for the fault forms which involve multiple generation units, they cannot be effectively reflected. By introducing the idea of hierarchical protection, the station layer protection is added on the basis of the original local layer protection. In this way, the protection can adaptively determine the action characteristic and tripping strategy according to the system operating situations and other generators' fault conditions. Through the coordination of the station layer, the local protection performance, the system control mode and the protection security defense ability can be advanced. In this section, the hierarchical out-of-step protection is taken as an example to illustrate the hierarchical protection construction method among different protection layers based on multi-generators information fusion.

3.2.1. Review of the Existing Out-of-Step Protection

The three-component impedance characteristic out-of-step protection criterion is usually used for large generators, and the protection action strategy is determined according to the impedance trajectory variation law [29,30]. Existing out-of-step protection only uses the local information, lacking information interaction and protection cooperation between generators. In addition, the protection fixed value is usually set according to the single generator out-of-step fault condition shown in Figure 7. However, since the current large power plants usually adopt multi-generators connected at one bus structure to supply power to the external power grid, when multi-generators become out-of-step at the same time, there may be a variety of out-of-step fault forms. Under different out-of-step forms, the existing protection mode faces difficulties in the action characteristics and generator tripping strategies.

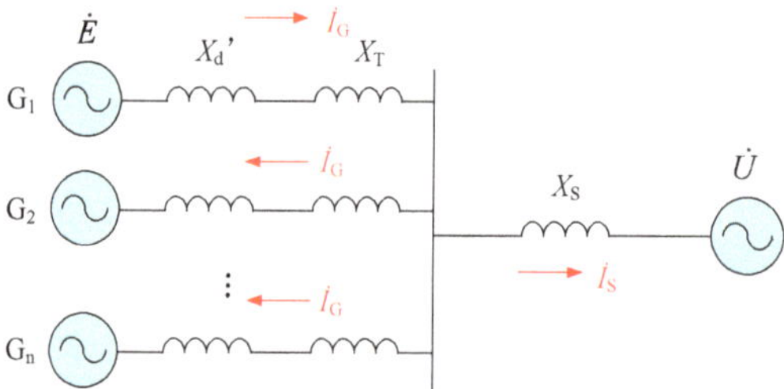

Figure 7. Schematic diagram of the single generator out-of-step condition.

As for the multi-generators connected at one bus out-of-step condition shown in Figure 8, for any out-of-step generator, the equivalent impedance of the system will be influenced by the oscillating current added from other generators. Thus, the measured impedance at the generator terminal will increase and its value is closely related to the operating mode of the power plant. Since the equivalent impedance value of the system is different in different out-of-step modes and different operating modes, the existing protection boundary cannot accurately identify the multi-generators out-of-step mode, which will result in protection maloperation or rejection.

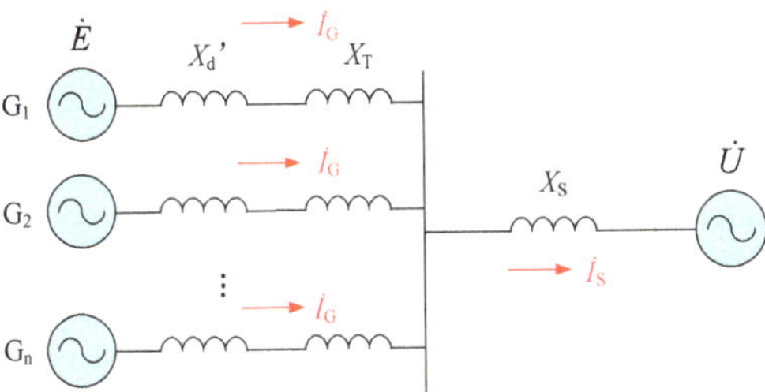

Figure 8. Schematic diagram of the multi-generators out-of-step condition.

In addition, for the existing out-of-step protection, each protection unit has the same protection fixed value. As long as the out-of-step oscillation between the generator and the external system is detected and the protection criterion meets action conditions, the protection will act and the corresponding generator will be tripped. However, this tripping strategy has no coordination with other connected generators. As for the multi-generators out-of-step condition, all the generators will be tripped at the same time, which harms the stability of the power system.

3.2.2. Hierarchical Out-of-Step Protection

In order to solve the protection principle defect and the action coordination problem under the multi-generators connected at one bus out-of-step conditions, combined with the existing three-component out-of-step protection criterion, the hierarchical out-of-step protection is constructed based on multi-generators information fusion. The out-of-step fault form can be identified by the cooperation of the station and local protection criterions, and the protection fixed value can be adaptively adjusted according to the operating mode of the power plant and the power system. The hierarchical out-of-step protection scheme is shown in Figure 9.

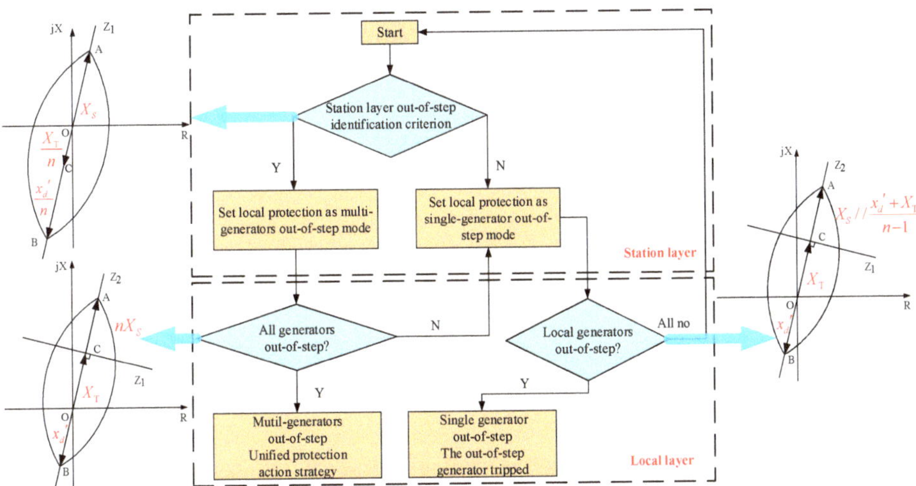

Figure 9. Hierarchical out-of-step protection.

The basic protection configuration, action characteristics and tripping strategy are set as follows:

(1)   The basic protection configuration. In order to detect the multi-generators out-of-step fault condition, the station layer out-of-step protection criterion is set at the

bus. The local out-of-step protection criterion of each generator is retained to detect the single generator out-of-step fault. Through information interaction with the station layer protection, the local out-of-step protection criterion can adapt to different out-of-step modes and operating conditions.

(2)   The protection action characteristics. The out-of-step protection action characteristics of the station layer and the local layer under different modes are shown in Figure 9. It is assumed that the parameters of all the generation units connected at the bus are the same. Among them, $x_d{}'$ means the subtransient reactance of the generator, X_T means the transformer reactance, X_S means the equivalent system reactance, n means the number of the generators operating in parallel. The out-of-step criterion of the station layer is set according to the impedance characteristics under multi-generators connected at one bus out-of-step fault condition. When it is judged that all the generators are out-of-step, the out-of-step criterion of the local layer is adaptively adjusted based on the enhancement effect of other generators. In the local layer, if the measured impedance trajectory of each local generator enters the multi-generators out-of-step action zone and meets the action condition, it can be ensured that the all the generators are out-of-step. Otherwise, the out-of-step criterion of the local layer should use the single generator out-of-step mode.

(3)   The protection tripping strategy. When the protection judging result is the single generator out-of-step condition, the out-of-step generator is directly tripped at the appropriate time. When the judging result is multi-generators all in the out-of-step condition, the instruction of the generators tripping by turns is issued by the station layer and executed by the local layer. Through information interaction with the grid-side, the tripping instructions are coordinated with the system stability control. After each generator is tripped, the station layer and local layer out-of-step protection need to measure the impedance trajectory again to determine the current out-of-step condition and tripping strategy. Until the system stability is restored, the protection returns.

3.2.3. Simulation Analysis of Protection Action Conditions under Multi-Generators Out-of-Step Fault

In order to verify the effectiveness of the hierarchical out-of-step protection scheme, the equivalent model shown in Figure 10 is constructed based on PSCAD/EMTDC software. Figure 10 shows that in large hydropower plant 1, 1# generator–transformer group and 2# generator–transformer group operate in parallel at the same bus M1. They are connected to the infinite system S through the 500 kV double-loop AC lines with the length of 200 km. Among them, G2 and T2 are formed by equivalent generator–transformer units. They are under the same working conditions and with the same capacity. The actual number of the generator–transformer units depends on the specific simulation condition, and the maximum is three. In hydropower plant 2, G3 and T3 are formed by equivalent generator–transformer units with four parallel operating generator–transformer units. All the generator–transformer units in the model are equipped with prime mover, governor and excitation systems. The model can reproduce the operation condition and control process of the actual multi-generators system to some extent and meet the basic requirements of the multi-generators out-of-step fault simulation.

Four generator–transformer units are set to operate in parallel at bus M1. Before the fault occurs, each generator sends out the active power as P = 600 MW and reactive power as Q = 167 Mvar. Four generator-transformer units are set to operate in parallel at bus M2. Before the fault occurs, each generator sends out the active power as P = 400 MW and reactive power as Q = 116 Mvar. The three-phase short-circuit fault is set at 0.5 s on Line1 II near the bus M1, and the fault line is removed at 0.61 s by protection action. The power angle changes of the three equivalent generators are shown in Figure 11.

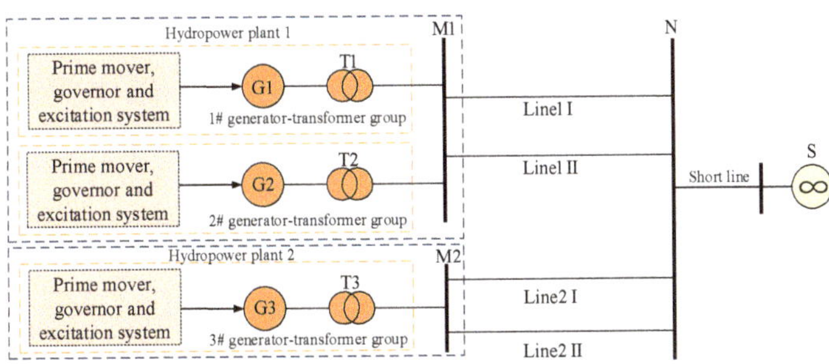

Figure 10. Schematic diagram of the out-of-step fault simulation model system.

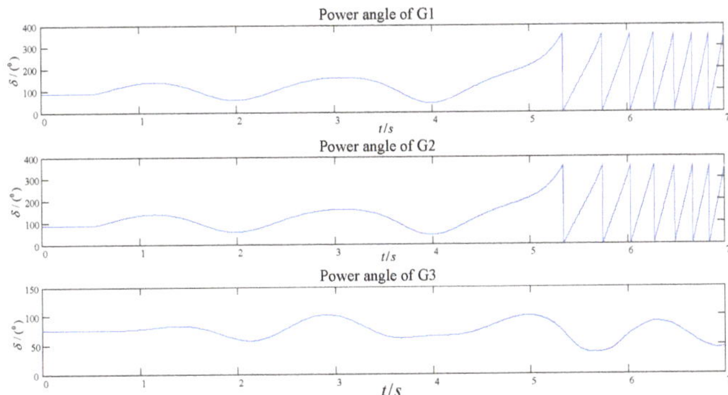

Figure 11. Power angle curve of the equivalent generators.

In Figure 11, after the fault is removed for a period of time, the equivalent generators G1 and G2 are out-of-step, and the equivalent generator G3 is oscillating synchronously. Therefore, the equivalent generators G1 and G2 constitute the situation of multi-generators connected at one bus out-of-step fault, and the equivalent generator G3 wobbles. According to the variation of the measured impedance trajectory at each bus during the simulation period, the action characteristics of the station layer out-of-step protection are shown in Figure 12.

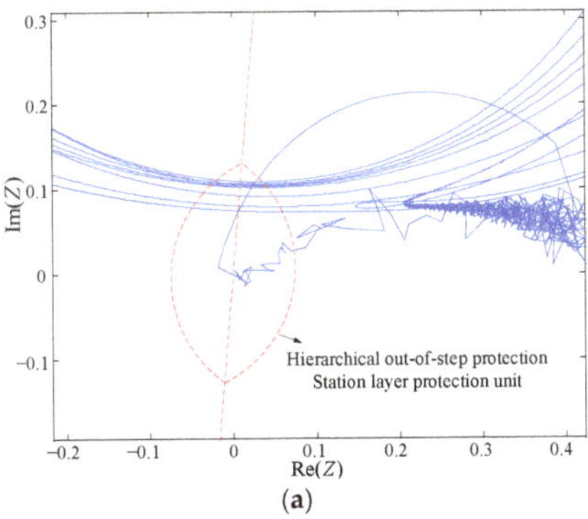

Figure 12. *Cont.*

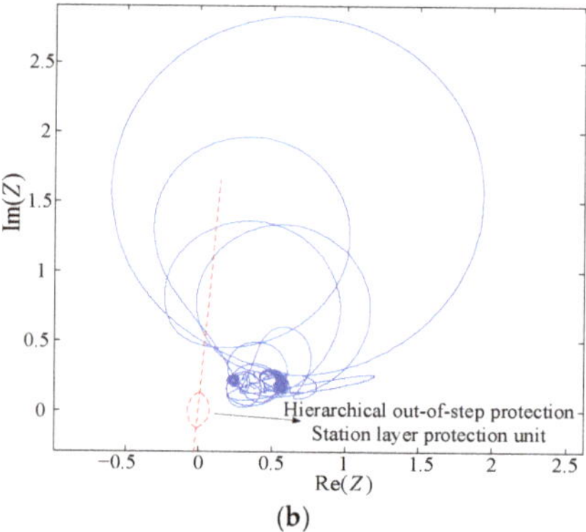

(b)

Figure 12. Measured impedance trajectory and the station layer protection action characteristics at each bus: (**a**) measured impedance trajectory at M1 and the station layer out-of-step protection action characteristics and (**b**) measured impedance trajectory at M2 and the station layer out-of-step protection action characteristics.

In Figure 12, the measured impedance trajectory at M1 can stably enter the action area of the station layer out-of-step protection. In order to verify the judgment of the station layer out-of-step protection, the local layer action areas of the generators G1 and G2 are defined as the multi-generators out-of-step mode. However, since the measured impedance trajectory at M2 does not enter the action area of the station layer protection, it can be judged that the equivalent generator G3 is only synchronous oscillation. Therefore, the station layer judges that there is not the multi-generators out-of-step fault at M2, so the local layer action area of G3 is defined according to the single generator out-of-step mode, which is the same as the existing out-of-step protection action area. The impedance trajectory measured at the terminal of each equivalent generator and the corresponding out-of-step protection action characteristics are shown in Figure 13.

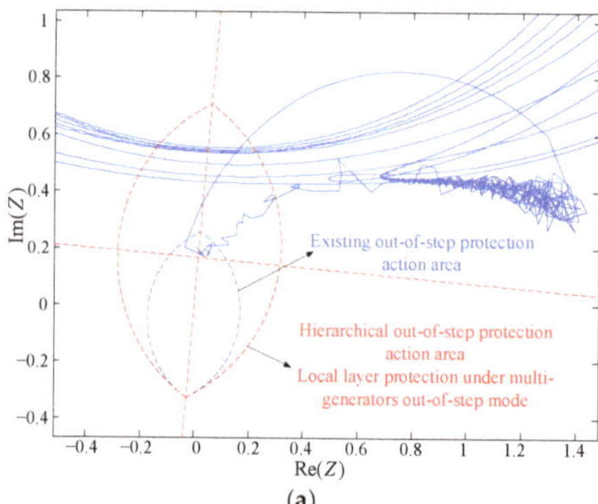

(a)

Figure 13. *Cont.*

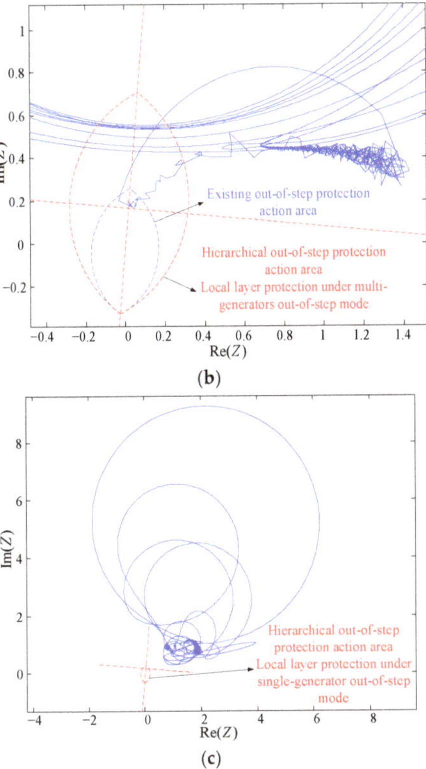

Figure 13. Measured impedance trajectory and the local layer protection action characteristics: (**a**) measured impedance trajectory of G1 and the local layer out-of-step protection action characteristics, (**b**) measured impedance trajectory of G2 and the local layer out-of-step protection action characteristics and (**c**) measured impedance trajectory of G3 and the local layer out-of-step protection action characteristics.

In Figure 13, the red dotted line represents the action characteristic of the proposed hierarchical local layer out-of-step protection and the blue dotted line represents the action characteristic of the existing out-of-step protection. According to the simulation results in Figure 13a,b, the measured impedance trajectory of the out-of-step generator only enters the action area of the existing out-of-step protection action characteristic within a short time, and the impedance trajectory of the stable out-of-step oscillation is completely outside the action area. Thus, the existing out-of-step protection refuses to act. Since the hierarchical out-of-step protection has considered the increasing effect of the oscillating current on the system impedance consisted by other generators, the protection action area has been adjusted. The simulation results in Figure 13a,b show that the multi-generators connected at one bus out-of-step fault can be accurately identified, and the crossing time of the measured impedance trajectory meets the requirements. In Figure 13c, since station layer has judged that the equivalent generator G3 is only synchronous oscillation, the protection action characteristics of the hierarchical local layer out-of-step protection and the existing out-of-step protection are the same. The simulation results show that the measured impedance trajectory does not enter the action area, and the protection will not act.

Hierarchical out-of-step protection can realize the coordination between the station and local layer protection through real-time information interaction between different layers and can effectively solve the problem that the existing protection scheme cannot accomplish the identification of multi-generators connected in one bus out-of-step fault conditions. The protection area boundary can be adaptively adjusted according to the actual operation mode of the system and the power plant, so that the protection criterion can adapt to the real-time operating condition. In addition, by fusing the protection information from the stability control system of the grid-side, a reasonable generator tripping strategy can be constructed to guarantee the safe operation of the generators and the power grid.

4. Key Problems of the Hierarchical Power Generation-Side Protection System

There are many problems which need to be studied to construct a complete multi-information fusion-based hierarchical power generation-side protection system. The key problems include the following aspects shown in Figure 14.

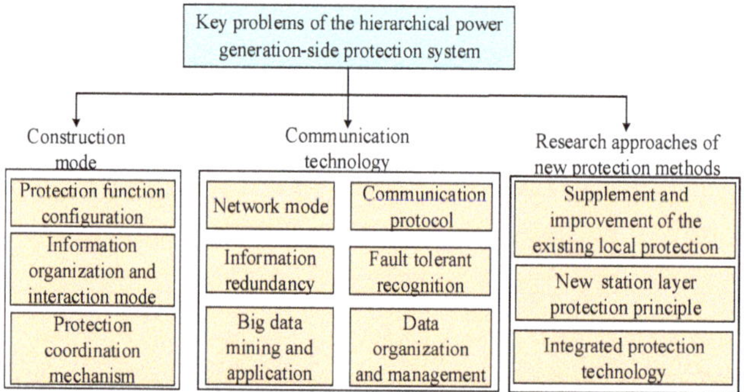

Figure 14. Key problems of the hierarchical power generation-side protection system.

4.1. Construction Mode of the Hierarchical Power Generation-Side Protection System

The local and station layer protection function should be determined based on the security and coordinated operation requirements of the protection performance and the information acquisition requirements of the hierarchical protection system. In order to determine the information organization and interaction mode among the hierarchical protection layers between generation units and between the generation and grid sides, the information from the CMS and the protection layers need to be analyzed. Moreover, the coordination mechanism between the station layer protection and the local layer protection needs to be studied.

4.2. Communication Technology of the Hierarchical Power Generation-Side Protection System

According to the data interaction requirements of the hierarchical protection system, the communication network mode and the data protocol need to be designed. In addition, the communication interaction method between the system generation and grid sides needs to be determined. The information redundancy and fault tolerant identification technology need to be studied to analyze the reliability of the communication system. Furthermore, the big data technologies need to be studied such as information mining and application, data organization and management, etc.

4.3. Research Approaches of New Hierarchical Power Generation-Side Protection Methods

The multi-information fusion-based hierarchical protection system opens a new way to improve the power generation-side protection. The research of the new hierarchical protection power generation-side protection methods can be started from the following aspects:

(1) Research on the supplement and improvement of the existing protection principle. This includes the improvement of the existing local protection and grid-related protection. The improved protection principle should fully adapt to the protection requirements under the complex power network operating environment.

(2) Research on the station layer protection based on the collaborative operation between the power generation and grid sides. This kind of research is rich in content, including the safe cooperated operation between the generation and grid sides, and making full use of the sufficient information support from the hierarchical protection system, etc. The protection should attain the purpose of serving the safe and stable operation of the power system, so as to ensure the cooperation and interaction between the two system sides. On the premise of meeting the healthy condition of the generation-side units and the regulation demand of the power grid, the effective coordination of the protection action strategies should be realized.

(3) Research on comprehensive protection technology to ensure the efficient operation of the power generation and grid sides. Based on multi-information sources, the health conditions of the generation units and the safety restrictions of the power grid can be reflected. Through reasonable cooperation between the hierarchical layers, the generation control should be accomplished under real-time security constraints.

5. Conclusions

In order to improve the generation-side security and the coordination performance between the power system generation and grid sides, the existing power generation-side protection mode is improved through the idea of a multi-information fusion-based hierarchical protection system.

(1) This paper presents a multi-information fusion-based hierarchical generation-side protection system. The system layer provides the system dispatch and control information to enable the generation-side to master the operation condition and dispatching demand of the system. The station layer realizes the information interaction among the generation units, the power grid and the automatic monitoring system. The local layer integrates abundant protection and control information to improve the protection performance.

(2) In view of the existing protection problems, hierarchical protection methods are constructed based on the multi-information fusion-based comprehensive local layer protection method and the multi-generators information fusion-based hierarchical protection method. Case analysis proved that, compared with the existing protection methods, the hierarchical single transverse differential protection comprehensive criterion can effectively enhance the protection sensitivity under the generator internal faults. The hierarchical out-of-step protection can adaptively adjust the protection action characteristic and generator tripping strategy according to the system operating condition and other generators' fault conditions, which can provide reliable guarantee for the safe operation of the generation and grid sides.

(3) The proposed protection methods are built on the basis of a perfect communication network, and the effect of information distortion is not considered. In order to provide reference and inspiration for the follow-up research, this paper discusses the key problems of the multi-information fusion-based hierarchical generation-side protection system which need to be further studied.

Author Contributions: Conceptualization, X.Y. (Xianggen Yin), Y.W. and J.Q.; data curation, X.Y. (Xianggen Yin), Y.W., J.Q. and X.Y. (Xin Yin); Formal analysis, X.Y. (Xianggen Yin); funding acquisition, X.Y. (Xianggen Yin); investigation, Y.W., J.Q., W.X. (Wen Xu), X.Y. (Xin Yin) and L.J.; methodology, X.Y. (Xianggen Yin) and Y.W.; resources, X.Y. (Xianggen Yin), Y.W., W.X. (Wen Xu), X.Y. (Xin Yin) and L.J.; validation, J.Q. and W.X. (Wen Xu); writing—original draft, X.Y. (Xianggen Yin), Y.W. and J.Q.; writing—review and editing, Y.W., W.X. (Wen Xu), L.J. and W.X. (Wei Xi). All authors have read and agreed to the published version of the manuscript.

Funding: This work was supported by the key project of smart grid technology and equipment of national key research and development plan of China under Grant 2016YFB0900600.

Institutional Review Board Statement: Not applicable.

Informed Consent Statement: Not applicable.

Data Availability Statement: Data is contained within the article.

Conflicts of Interest: The authors declare no conflict of interest.

References

1. Technical Report on the Events of 9 August 2019. Available online: https://www.nationalgrideso.com/document/152346/download (accessed on 1 August 2020).
2. Appendices to the Technical Report on the Events of 9 August 2019. Available online: https://www.nationalgrideso.com/document/152351/download (accessed on 5 August 2020).
3. Huadong, S.; Tao, X.; Qiang, G.; Yalou, L.; Weifang, L.; Jun, Y.; Wenfeng, L. Analysis on Blackout in Great Britain Power Grid on 9 August 2019 and Its Enlightenment to Power Grid in China. *Proc. CSEE* **2019**, *39*, 6183–6192.
4. Project Group Turkey. *Report on Blackout in Turkey on 31 March 2015*; ENTSO-E: Brussels, Belgium, 2015.
5. Yao, S.; Yong, T.; Jun, Y.; Ansi, W. Analysis and lessons of Blackout in Turkey Power Grid on 31 March 2015. *Autom. Electr. Power Syst.* **2016**, *40*, 9–14.
6. Weifang, L.; Jun, Y.; Qiang, G.; Zhiwen, W.; Qi, J.; Fangfang, Y. Analysis on Blackout in Argentine Power Grid on 16 June 2019 and Its Enlightenment to Power Grid in China. *Proc. CSEE* **2020**, *40*, 2835–2842.
7. Jiandong, Y.; Kun, Z.; Ling, L.; Peng, W. Analysis on the causes of units 7 and 9 accidents at Sayano-Shushenskaya hydropower station. *J. Hydroelectr. Eng.* **2011**, *4*, 226–234.
8. Razzaghi, R.; Davarpanah, M.; Sanaye-Pasand, M. A Novel Protective Scheme to Protect Small-Scale Synchronous Generators Against Transient Instability. *IEEE Trans. Ind. Electron.* **2013**, *60*, 1659–1667. [CrossRef]
9. Paudyal, S.; Ramakrishna, G.; Sachdev, M.S. Application of Equal Area Criterion Conditions in the Time Domain for Out-of-Step Protection. *IEEE Trans. Power Deliv.* **2010**, *25*, 600–609. [CrossRef]
10. Wang, Y.; Yin, X.; Zhang, Z. The Fault-Current-Based Protection Scheme and Location Algorithm for Stator Ground Fault of a Large Generato. *IEEE Trans. Energy Convers.* **2013**, *28*, 871–879. [CrossRef]
11. Guilin, L.; Wei, S.; Xinli, S. Research on coordination of low excitation limit, loss of excitation protection and out-of-step protection based on grid-related protection. *Power Syst. Prot. Control* **2014**, *42*, 107–112.
12. Yang, Q.; Tang, X.; Su, Y.; Xiang, L.; Ding, M.; Mei, Y. An Optimal Allocation Scheme for Relay Setting Related to Power Grid and Over Frequency Generator Tripping Measures. *Autom. Electr. Power Syst.* **2013**, *37*, 127–131+135.
13. Phadke, A.G. Synchronized phasor measurements in power systems. *IEEE Comput. Appl. Power* **1993**, *6*, 10–15. [CrossRef]
14. Lee, G.; Kim, D.-I.; Kim, S.H.; Shin, Y.-J. Multiscale PMU Data Compression via Density-Based WAMS Clustering Analysis. *Energies* **2019**, *12*, 617. [CrossRef]
15. Bagher, M.; Jabali, A.; Kazemi, M.H. Power System Event Ranking Using a New Linear Parameter-Varying Modeling with a Wide Area Measurement System-Based Approach. *Energies* **2017**, *10*, 1088.
16. Serizawa, Y.; Myoujin, M.; Kitamura, K.; Sugaya, N.; Hori, M.; Takeuchi, A.; Shuto, I.; Inukai, M. Wide-area current differential backup protection employing broadband communications and time transfer systems. *IEEE Trans. Power Deliv.* **1998**, *13*, 1046–1052. [CrossRef]
17. Serizawa, Y.; Imamura, H.; Kitamura, K. Performance evaluation of IP based relay communications for wide area protection employing external time synchronization. In Proceedings of the 2001 IEEE Power Engineering Society Summer Meeting, Vancouver, BC, CA, 15–19 July 2001; pp. 909–914.
18. Tan, J.C.; Crossley, P.A.; Kirschen, D.; Goody, J.; Downes, J.A. An expert system for the back-up protection of a transmission network. *IEEE Trans. Power Deliv.* **2000**, *15*, 508–514. [CrossRef]
19. Shalini; Samantaray, S.R.; Sharma, A. Enhancing Performance of Wide-Area Back-Up Protection Scheme Using PMU Assisted Dynamic State Estimator. *IEEE Trans. Smart Grid* **2019**, *10*, 5066–5074. [CrossRef]
20. Jena, M.K.; Samantaray, S.R.; Panigrahi, B.K. A New Adaptive Dependability-Security Approach to Enhance Wide Area Back-Up Protection of Transmission System. *IEEE Trans. Smart Grid* **2018**, *9*, 6378–6386. [CrossRef]
21. He, J.; Zhu, G.; Bo, Z. Integrated Protection for Power Systems Based on the Multi-Agent Technology. *Trans. China Electrotech. Soc.* **2007**, *6*, 141–147.
22. Bo, Z.Q.; He, J.H.; Dong, X.Z.; Caunce, B.R.J.; Klimek, A. A multi-algorithm based integrated protection scheme for distribution systems. In Proceedings of the 7th Institution of Engineering and Technology International Conference on Advances in Power System Control, Operation and Management (APSCOM 2006), Hong Kong, China, 30 October–2 November 2006.
23. Weijian, W. *Electrical Equipments Relaying Protection Principle and Application*, 2nd ed.; Electric Power Press: Beijing, China, 2001.
24. Cui, X.; Tai, N.; Liu, J. New Unit-transverse Differential Protection Based on the Fault Contributed Zero Sequence Current. *Autom. Electr. Power Syst.* **2007**, *12*, 61–63.
25. Zhang, K. *Protection System and New Dynamic Simulation Experiment Technology of Huge Hydro Generator*; Huazhong University of Science and Technology: Wuhan, China, 2008.
26. Zhang, K.; Yin, X.; Chen, D.; Zhang, Z. Dynamic Simulation Experiments for Internal Fault Protection of Large Hydro-generators. *Autom. Electr. Power Syst.* **2008**, *6*, 85–90.

27. Yang, J. *Study on Fault Transient Simulation for Huge Hydro-Generators and Protection Technology for Generator-Transformer Units*; Huazhong University of Science and Technology: Wuhan, China, 2004.
28. Xia, Y. *Study and Application on Huge Hydro Generator Internal Fault Transient Simulation and Optimized Main Protection Scheme*; Huazhong University of Science and Technology: Wuhan, China, 2006.
29. Anderson, P.M. *Protective Schemes for Stability Enhancement, in Power System Protection*; IEEE Press: New York, NY, USA, 1998.
30. Reimert, D. *"Loss of Synchronism" Protective Relaying for Power Generation Systems*; CRC Press Taylor & Francis Group: New York, NY, USA, 2006.

Article

Protection Method Based on Wavelet Entropy for MMC-HVDC Overhead Transmission Lines

Weibo Huang, Guomin Luo *, Mengxiao Cheng, Jinghan He, Zhao Liu and Yuhong Zhao

School of Electrical Engineering, Beijing Jiaotong University, Beijing 100044, China; 15117384@bjtu.edu.cn (W.H.); 19121415@bjtu.edu.cn (M.C.); jhhe@bjtu.edu.cn (J.H.); 98930225@bjtu.edu.cn (Z.L.); 19126205@bjtu.edu.cn (Y.Z.)
* Correspondence: gmluo@bjtu.edu.cn or guominluo@hotmail.com; Tel.: +86-10-5168-3740

Abstract: Recent technological developments in modular multilevel converter-based high-voltage direct current (MMC-HVDC) transmission systems have shown significant advantages over the traditional HVDC and two-level voltage source converter (VSC) transmission systems. However, there are a lack of studies on the protection methods for MMC-HVDC overhead lines where the protection method should be able to provide a fast and accurate response and be able to identify lightning strikes. In this paper, a wavelet entropy-based protection method is proposed. Due to the capability of revealing time–frequency distribution features, the proposed protection method combines wavelet and entropy to identify the time–frequency characteristics of different faults. Simulation results show that the proposed method can accurately and quickly determine the types of faults or disturbances with appropriate noise tolerance. In addition, the impact of the ground resistor and fault distance on the performance of the proposed method is studied.

Keywords: wavelet entropy; transient component; MMC-HVDC; protection

---

Citation: Huang, W.; Luo, G.; Cheng, M.; He, J.; Liu, Z.; Zhao, Y. Protection Method Based on Wavelet Entropy for MMC-HVDC Overhead Transmission Lines. *Energies* **2021**, *14*, 678. https://doi.org/10.3390/en14030678

Academic Editor: Adam Dyśko

Received: 9 November 2020
Accepted: 27 January 2021
Published: 28 January 2021

Publisher's Note: MDPI stays neutral with regard to jurisdictional claims in published maps and institutional affiliations.

1. Introduction

In the last two decades, considerable progress has been made in the development of voltage source converter-based high-voltage direct current (VSC-HVDC) transmission systems. It is regarded as a promising solution for controlling real and reactive power and reducing losses. Among the different design topologies of VSCs, the modular multilevel converter (MMC) has many advantages, such as modular design, good scalability, low switching frequency, and few harmonic injections [1–3]. It serves as an effective solution for the large capacity and long-distance transmission systems which connect large-scale renewable energy resource centers in western China with the load centers on the eastern coast of China. At the same time, it improves the flexibility and reliability of the power grid operations [4,5]. Comparing with the traditional line-commuted converters (LCCs), VSC-based converters introduce completely different fault characteristics [6]. Especially, for a MMC-based HVDC system, fast and reliable protection is required to eliminate huge fault currents and discriminate transient disturbances. As a result, the protection methods of the traditional HVDC cannot be applied to MMC-HVDC systems. It is necessary to study the fault characteristics of the MMC and propose a protection method for the MMC-based transmission lines.

In traditional LCC-based HVDC transmission systems, the fault current is fully controllable, so line protection does not need to be very fast [7,8]. In the symmetrical monopole MMC-HVDC system, when a pole-to-pole fault (PPF) occurs, the bridge arm sub-module (SM) has a discharge circuit, which causes the fault current usually to climb to tens of times the rated current in a few milliseconds [9,10], and the AC power will feed to the fault point through the freewheeling diode. That is, the fault current is uncontrollable. Therefore, it is necessary to detect and isolate DC faults as fast as possible, such as in 3 ms [11,12]. In addition, when a pole-to-ground fault (PGF) occurs, the bridge arm SM has no discharge circuit. It will only cause small current fluctuations during the transient

149

period, which cannot easily be detected. Furthermore, the transient characteristics caused by lightning faults and lightning disturbances are also slight fluctuations in the current. Their characteristics are very similar and need to be effectively distinguished.

In existing LCC-based HVDC projects, the time of traveling wave protection is about 10 ms, and the action time of differential under-voltage protection is about 20 ms [13–16]. Compared with traditional HVDC, when a PPF occurs in MMC-HVDC, the fault current rises rapidly, which requires higher protection speed and sampling frequency. In this case, the reliability and sensitivity of traveling wave protection and differential under-voltage protection will be reduced. In addition, in terms of the transient recognition of lightning strikes, traditional LCC-based HVDC protection methods mainly distinguish between lightning and faults from the perspective of frequency domain spectrum characteristics or time domain waveform characteristics. However, under short time window scenarios, the resolutions in the frequency domain and discriminations in the time domain waveform are both reduced. Therefore, the protection method of the traditional LCC-based HVDC system does not work for the MMC-HVDC system. New requirements are put forward for the identification of interference from lightning strikes and noise. Compared with the traditional LCC-based HVDC system, the protection method of the MMC-HVDC system needs to consider the reliability of protection and the accuracy of lightning identification at extremely fast speeds.

In recent years, research on the protection of MMC-HVDC transmission lines mainly includes protection methods based on sudden changes of voltage or current protection [17,18], pilot protection [19–21], and traveling wave protection and its improvement methods [22–24]. Among them, the sudden change in voltage or current protection is based on the protection principle of single-ended electrical information. It has extremely fast action speed and is also sensitive. The pilot protection improves the reliability of the protection through the comparison and processing of double-ended information. Traveling wave protection uses corresponding mathematical methods to extract and analyze the high-frequency characteristics of transient traveling waves, such as amplitude, polarity, duration, etc., which has an extremely fast speed for fault detection. However, existing studies still have the following shortcomings: firstly, some protection methods do not consider the impact of lightning disturbances; secondly, some protection methods adopt double-ended information, which might not meet the requirements of the MMC-HVDC system for primary protection speed; finally, some protection methods need to be strengthened in terms of tolerance to ground resistor and anti-interference ability. In general, the protection method of MMC-HVDC is not yet mature and needs further study. Therefore, this article pays more attention to how to improve the reliability of protection and the ability to identify lightning strikes and interference when the protection satisfies high-speed mobility.

In MMC-HVDC systems, different faults have different transient characteristics and different time–frequency distributions. Some faults have similar time domain waveforms, but their frequency domain distributions are different. Therefore, it is possible to propose a protection method based on the difference in time–frequency distribution. The existing signal analysis tools include Fourier transform, Hilbert–Huang transform, S transform, etc. They are the mainstream methods in the field of power system relay protection. However, the amplitude and ratio are usually used as the classification criteria, and consider fewer influencing factors. When considering the influence of factors such as noise and lightning, its setting calculation is complicated, and there are many simplified and equivalent treatments. Its stability and reliability need to be improved. Wavelet transform, a time–frequency analysis method, has the characteristics of multi-resolution analysis, and a strong ability to represent information [13,25,26]. Entropy is a tool to measure the disorder degree of the whole system [27–29]. Wavelet entropy is a combination of wavelet transform and information entropy. It can not only achieve the purpose of information fusion but also analyze the mutation signal more effectively [30,31]. It has been gradually applied in image processing, stock market forecasting, biology, machinery, power systems, and other fields, with good achievements [32,33]. It has advantages in characterizing system distribution characteristics and degree of disorder.

In this paper, wavelet entropy is used to represent the characteristics of different fault transient characteristics, which can not only meet the requirements of the MMC-HVDC system for protection speed, but also has a good ability to distinguish lightning strikes. The main contributions of the paper are the following:

1. The characteristics of different faults and disturbances are detailed analyzed;
2. A protection method based on wavelet entropy for MMC-HVDC overhead transmission lines is proposed;
3. It is verified that the protection method is not affected by the distance of the fault, has a strong ability to withstand ground resistors, and has a strong noise tolerance.

The content of this paper from Sections 2–6 is as follows: Section 2 analyzes the MMC-HVDC two-terminal transmission model in detail and introduces the fundamentals of the MMC-HVDC system and characteristics of the different faults. Section 3 explains the definition of wavelet entropy and the principles of wavelet entropy in describing spectrum features. Section 4 proposes the methods of transmission line protection. Section 5 shows the simulation results and verifies the effectiveness of the protection method based on wavelet entropy. Section 6 offers conclusions and discussions.

2. Fault Characteristics of MMC-HVDC Transmission Lines

2.1. Fundamentals of MMC-HVDC System

Limited by the compactness, lightness, and capital cost requirements of the converter station, the symmetrical monopole MMC-HVDC system has a good application prospect. Both domestic and many foreign MMC-HVDC projects have adopted symmetrical monopole converter stations [34,35]. The research in this paper is based on a symmetrical monopole MMC-HVDC system, as shown in Figure 1. The converter stations serve as an interface between DC and AC systems. The half-bridge SM is commonly used in recent MMC projects [36,37]. The work of this paper is based on half-bridge MMC, and the number of SMs of each bridge arm is n. The DC side of the converter is grounded through a large resistor R_g and the neutral point is grounded. The bridge arm reactor L_0 can inhibit the interphase circulation current, reduce the bridge arm current's harmonic distortion rate, and suppress the fault current to protect equipment. L_m is a current-limiting reactor, which can effectively reduce the rate of change of the fault current after the DC line fault and leave more time for the protection of the DC line and the action of the DC circuit breaker. The DC current I_{dc} in the transmission line is measured at the point between the transmission line and the current-limiting reactor. In this paper, the transient states to be considered for protection include PPFs located at F1, PGFs, lightning disturbances (LDs), and lightning faults (LFs) located at F2, SMs short circuit faults (SMFs) on the arm of a phase located at F3, and single-phase grounding AC faults (AG-ACs) located at F4.

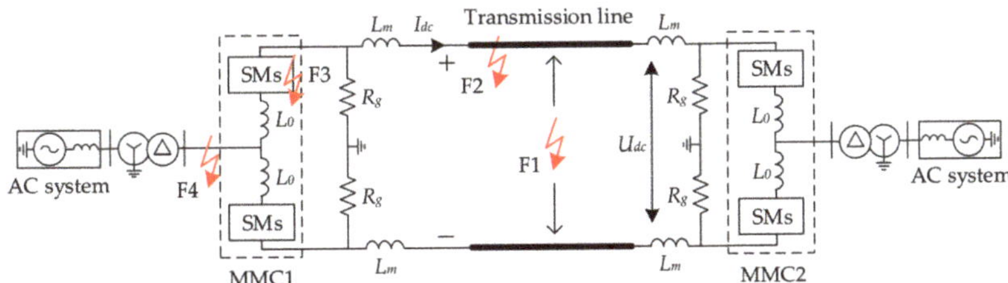

Figure 1. Diagram of a modular multilevel converter-based high-voltage direct current (MMC-HVDC).

Figure 2 shows the topology of a typical MMC converter, which has three identical phase units with six arms in total. Each arm consists of a reactor L_0 and n SMs in series.

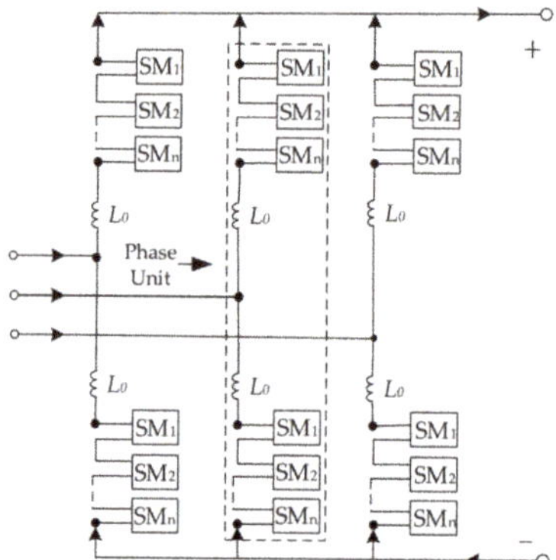

Figure 2. Topological structure of an MMC.

2.2. Characteristics Analysis of PPF

When a PPF occurs in an MMC-HVDC system, which is located at F1 in Figure 1, the fault process can be divided into three stages: the unblocked SM stage, the initial stage after the SM is blocked, and the steady stage after the SM is blocked. In order to detect a fault in a short time, it is necessary to use transient information at the beginning of the fault. Therefore, the fault transient process studied in this paper mainly focuses on the unblocked SM stage. The equivalent circuit of the unblocked SM stage is shown in Figure 3. The fault current I_{dc} is mainly composed of the SM capacitor discharging current and AC source feeding the current. During this stage, SMs are switched to the normal operation mode, which means a total of n SMs of each phase unit are in the on-state at any time to maintain the DC bus voltage. Due to the SM capacitance voltage balance control principle, SMs are switched on and off sequentially in a short time frame. All SMs in each phase can be divided into two groups, and discharge alternately, as shown in Figure 3b. Because of the high control frequency, it can be approximated that the two groups of SMs alternately discharged are in parallel in each phase. Therefore, it can be equivalent to a second order RLC discharging circuit, as shown in Figure 3c.

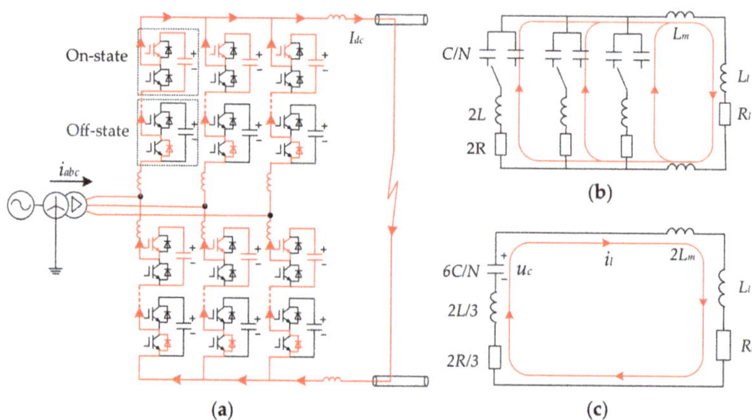

Figure 3. MMC-HVDC system pole-to-pole fault (PPF) circuit. (**a**) Detailed fault circuit; (**b**) equivalent fault circuit; (**c**) simplified fault circuit.

The waveform and frequency spectrum of fault current I_{dc} is shown in Figure 4. As shown in Figure 4a, the fault time is 0.01 s, and the fault current rises to around 14 kA after

10 ms. A fast isolation is needed to avoid the large-amplitude fault currents. If DC circuit breakers are used, the post-fault transients of only a few microseconds will be recorded. The maximum breaking capacity of existing DC circuit breakers is 25 kA. According to the breaking capability and the operation time of the DC breaker, there is only a 3 ms transient signal after the fault is studied in this article. Due to the coupling between the two transmission lines, the Karenbauer phase mode transformation matrix is used to decouple the currents in both positive and negative poles. When compared with the ground mode component, the line mode component of currents is relatively stable. Its wave speed varies relatively little with the frequency and geographic environment of the corridor. Thus, the line mode component is adopted. The waveform of the current line mode component is shown in Figure 4b. As shown in Figure 4c, the line mode component of the fault current is used for spectrum analysis. It can be seen that the amplitude of the 0 Hz component exceeds 5 kA. The content of each frequency band gradually decreases with increasing frequency. There are obvious differences between the content of different frequencies, both in the low-frequency band and the intermediate frequency band. However, from around 20 kHz to around 50 kHz, the frequency spectrum is evenly distributed and varies slightly.

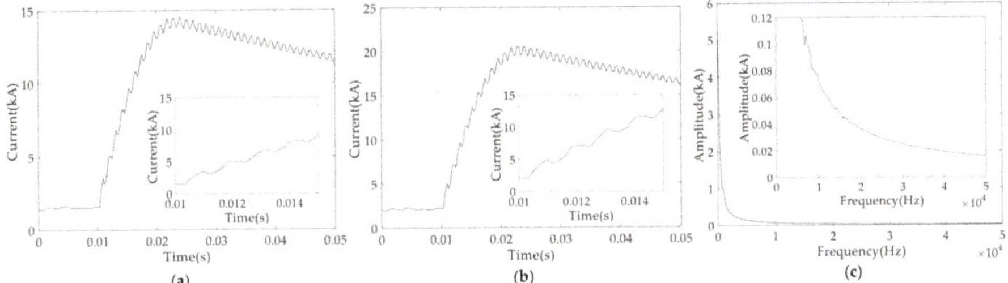

Figure 4. MMC-HVDC system PPF characteristics. (**a**) Current waveform; (**b**) current line mode component waveform; (**c**) spectrum analysis.

2.3. Characteristics Analysis of PGF

The characteristics of a PGF of MMC-HVDC are different from either traditional LCC-based HVDC converters or two-level VSC-based converters. When the PGF occurs at the location F2 in Figure 1, no closed loop can be formed between the faulted point and the SM capacitors. Only a slight transient signal is produced due to the deceasing bus potential of the faulted line. The capacitor will not discharge. Figure 5 illustrates the corresponding equivalent circuit. Therefore, the average SM capacitor voltage remains the same, and the DC voltage between two poles will be unchanged. The fault pole voltage becomes 0. There is no steady-state fault current in the DC line.

The PGF transients also have wave processes similar to the PPF. The difference is that a PGF forms a fault loop with the earth, which causes the transient wave to be severely attenuated during propagation. This transient process can be equivalent to the result of switching an additional voltage source at the fault point. The equivalent additional source E_{add} is:

$$E_{add}(t) = -E_{tra}\ (t \geq 0), \tag{1}$$

where E_{tra} represents the transient voltage source that generates the transient current waveform. The transient characteristics of the PGF current I_{dc} are shown in Figure 6. The energy decreases with the increase in frequency. More sizeable differences can be found in the frequency band from 0 Hz to around 10 kHz than the other bands. From around 10 to 50 kHz, the differences gradually decrease.

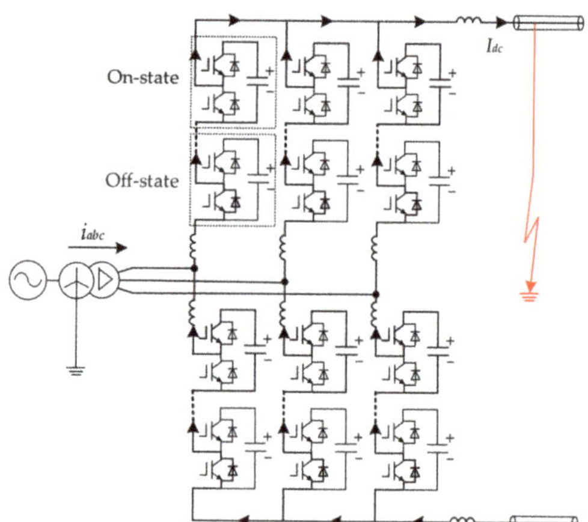

Figure 5. MMC-HVDC system pole-to-ground fault (PGF) circuit.

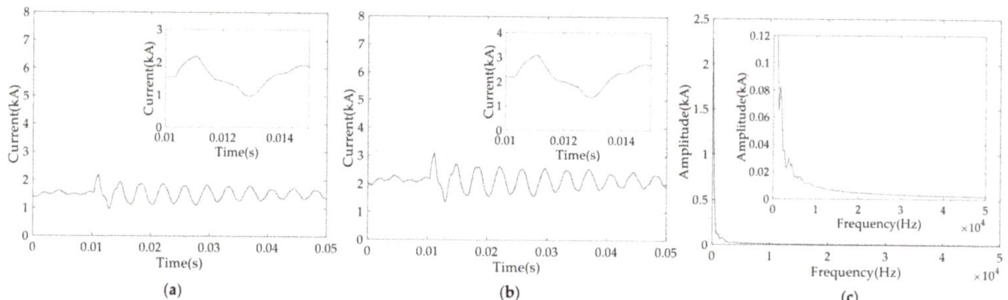

Figure 6. MMC-HVDC system PGF characteristics. (**a**) Current waveform; (**b**) current line mode component waveform; (**c**) spectrum analysis.

2.4. Influence Caused by Lightning Strikes

Lightning strikes are the main cause of transmission line protection misoperation. For the extremely short post fault transients of MMC-based protection, it is necessary to consider how to distinguish between faults and disturbances, especially LDs. Lightning waves generated by thunder are pulse transient waves, whose shape is mainly determined by its steepness and peak value. The double exponential wave is the equivalent calculation wave which is the closest to the actual lightning current wave, and is widely used in simulation analysis [38]. The generation of the transient waveform can be equivalent to superimposing an additional current source. The equation of the additional source E_{add} is:

$$E_{\text{add}}(t) = I_0(e^{-t/\alpha} - e^{-t/\beta})\,(t \geq 0),\tag{2}$$

Here, I_0 represents the amplitude of the lightning current; α and β represent the correlation coefficients of the rise and fall of lightning current, respectively.

When an LD occurs, which is located at F2 in Figure 1, it can be equivalent to a single current source superimposed on the transmission line. The signal source is connected for a short time and generates high-frequency signals. The transient characteristics of an LD fault current I_{dc} are shown in Figure 7. Most energy is contained in the lower frequency band, from 0 Hz to around 30 kHz, while only a small part of energy is included beyond 30 kHz. At the same time, large amplitude oscillations can be found in the frequency band below 15 kHz. Those harmonic energies cause the oscillating waveforms of lightning in the time domain.

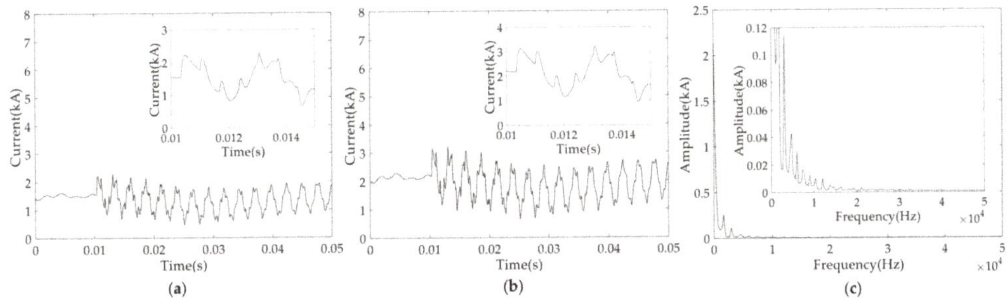

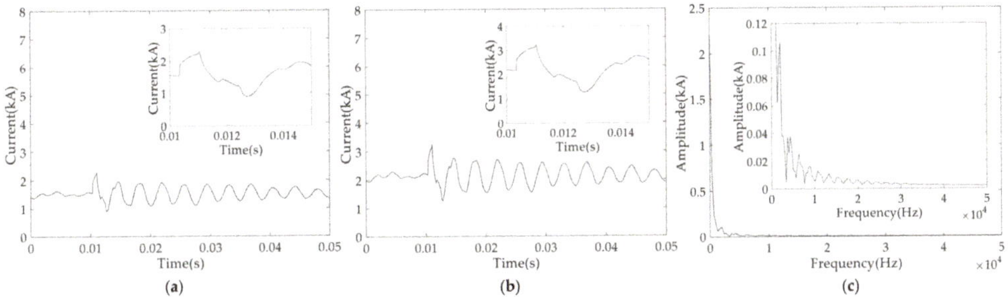

Figure 7. Transient characteristics of a lightning disturbance (LD). (**a**) Current waveform; (**b**) current line mode component waveform; (**c**) spectrum analysis.

When the lightning current amplitude is large, the voltage between the line and the tower might exceed the flashover voltage of the insulator. It is easy to cause an insulation breakdown, especially when the insulator is partially damaged, or flashover occurs along the surface. Then, it develops into a stable arc in a short time. The transmission line has a pole-to-ground fault through the tower, which is called a lightning fault (LF). According to its physical mechanism, the mathematical model of the additional source E_{add} can be expressed as a piecewise function in Equation (3),

$$E_{add}(t) = -E_{tra}\ (t_0 \le t \le \infty),\ \text{and}\ E_{add}(t) = I_0(e^{-t/\alpha} - e^{-t/\beta})\ (0 \le t \le \infty), \tag{3}$$

Here, t_0 is the moment of insulation breakdown, after which the LD evolves into an LF. The LF transient characteristics of I_{dc} are shown in Figure 8. The frequency spectrum of LF looks similar to that of LD in Figure 7c. However, they are different. Due to the superposition of GF, whose energy is focused in the low-frequency band, the energy spectrum of LF reveals the features of both GF and LD. As demonstrated in Figure 8c, most energy is concentrated in the range from 0 to 30 kHz. The energy decreases gradually beyond 30 kHz. Although more harmonic-like energies are included in the frequency from around 15 to 50 kHz, less steepness is shown in LF than that in LD.

Figure 8. Transient characteristics of a lightning fault (LF). (**a**) Current waveform; (**b**) current line mode component waveform; (**c**) spectrum analysis.

2.5. Characteristics Analysis of External Faults

When an external fault occurs, line protection should not be activated. In MMC-HVDC systems, external faults mainly include valve faults and AC side faults. Among them, SMFs and AG-ACs are the most likely faults, respectively. For that reason, these two faults are considered in this paper: an SMF located at F3 and an AG-AC located at F4 in Figure 1. Figures 9 and 10 show the fault transient characteristics of I_{dc} of the SMF and the AG-AC, respectively. For these two types of external faults, most of their energies are found between 0 Hz and around 5 kHz, but the energies attenuate greatly beyond 5 kHz.

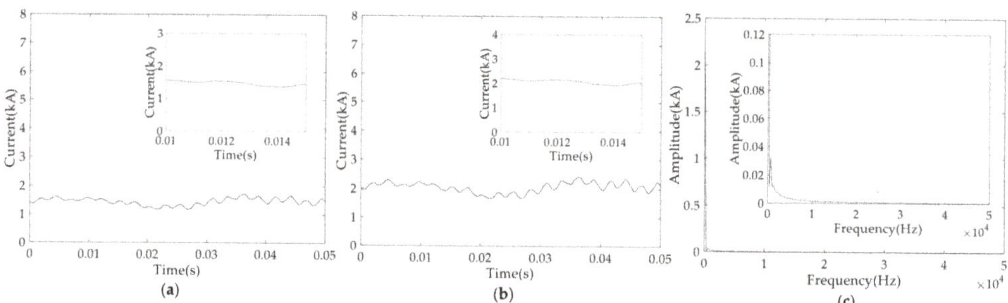

Figure 9. Fault characteristics of a sub-module short circuit fault (SMF) on the arm of a phase. (**a**) Current waveform; (**b**) current line mode component waveform; (**c**) spectrum analysis.

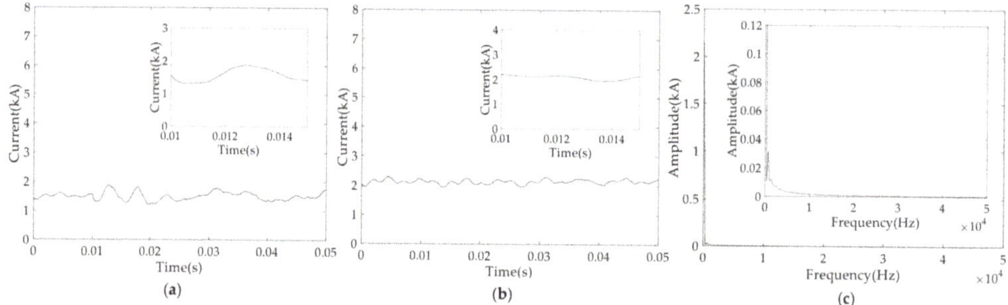

Figure 10. Fault characteristics of a single-phase grounding AC fault (AG-AC). (**a**) Current waveform; (**b**) current line mode component waveform; (**c**) spectrum analysis.

When an internal fault occurs, the transient current propagates from the fault point to both sides along the transmission line. Because the actual transmission line is basically a uniform line, which means that the resistor, inductance, and capacitance parameters are evenly distributed along the line, the high-frequency attenuations of the transient current at both ends of the line are not large. On the contrary, when an external fault occurs, the high-frequency component of the transient current attenuates significantly after passing through the boundary.

According to the analysis in this section, it can be found that different fault transients have different characteristics such as waveforms and frequency spectrum distributions. The high-frequency components of the transient current of the SMF and the AG-AC are obviously smaller than the ones of internal faults. The 0 Hz component of PPF is greater than those of PGF, LD, and LF. The spectra of PGF, LD, and LF in different frequency bands are obviously different, both the amplitude and the variations. If a reasonable signal processing and analyzing tool is used to describe these differences effectively and stably, various transients can be well separated, and the protection function can be realized.

3. Wavelet Entropy Characterization of Spectrum Distribution

3.1. Definition of Wavelet Entropy

Wavelet transform has good time–frequency localization performance [27,28]. It is considered as a "mathematical microscope", because it can "focus" the analysis object to any detail. Benefitting from the high sensitivity of the wavelets to the singularity and mutation of signals, wavelet transform is considered an effective signal processing method in multi-resolution analysis. Power system faults appear as sudden changes in voltage and current signals. Therefore, the use of a wavelet to detect the fault singularities is effective [29].

Entropy is one of the tools to measure the disorder degree of the whole system and it can also be regarded as a description of the degree of system uncertainty. If we regard a signal source as a material system, the more messages we may output and the more random and uncertain the signal source is, the more disordered and the greater the entropy is [13,25].

Wavelet entropy can represent the change in signal complexity in the time domain, and also many features of signal in the frequency domain. Therefore, wavelet entropy has unique advantages in the representation of fault information of non-stationary time-varying signals. According to wavelet transform, the entire frequency band is divided into m levels to obtain frequency bands of different levels. The wavelet coefficients in the ith ($1 \leq i \leq m$) frequency band form a set X_i [39,40]. Taking the maximum and minimum values of X_i as upper and lower limits, respectively, this range is equally divided into n intervals. The number of wavelet coefficients distributed in the jth ($1 \leq j \leq n$) interval is denoted as x_{ij}. Its probability is recorded as $p(x_{ij})$, which is obtained by dividing the number of coefficients in different intervals by the total number of coefficients in this frequency band X_i. The formula of wavelet entropy H_i of set X_i is:

$$H_i = -\sum p(x_{ij})log_b(x_{ij}) \ (j = 1, 2, \ldots , n),\tag{4}$$

3.2. Characterization of Frequency Spectrum by Wavelet Entropy

In MMC-HVDC systems, to ensure protection speed and sensitivity, the sampling frequency of the protection device is required to be no less than 50 kHz; a short time window requires a high sampling frequency to increase the acquisition of transient information. In most engineering applications, the sampling frequency is set as 100 kHz [17,18,20], and the largest is 1 MHz [22]. In this work, we choose 200 kHz as the sampling frequency. Then, the signal is decomposed by wavelet transform. Figure 11 shows the energy distribution of different wavelet frequency bands. Through spectrum analysis, there is a small amount of content of different frequencies from 50 to 100 kHz of various faults, and the change is insignificant. Therefore, the spectrum display range is from 0 to 50 kHz in this case. There are clear differences in the frequency spectrum of different faults in different frequency bands; in particular, the degrees of frequency fluctuation and disorder are different. For example, on the fourth level in the spectrum from 6250 Hz ($200 \text{ kHz}/2^{4+1}$) to 12,500 Hz ($200 \text{ kHz}/2^4$), the differences between the content of different frequencies of lightning strikes are larger than the PGF. These characteristics can be characterized by entropy.

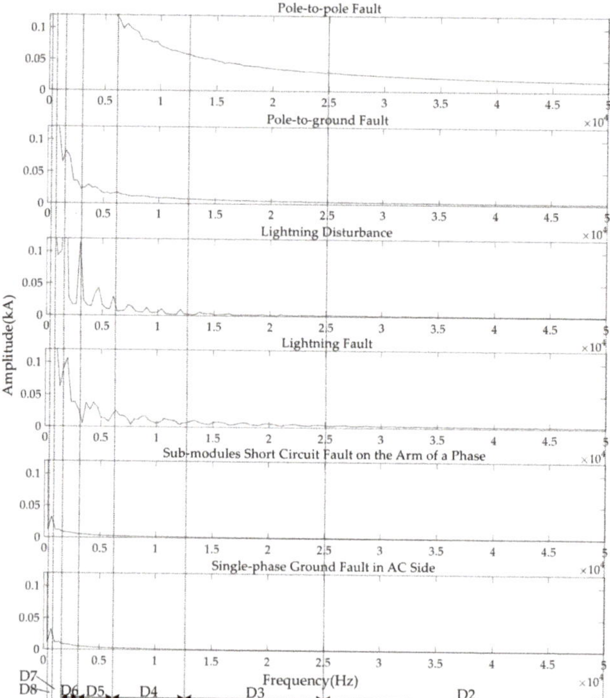

Figure 11. Spectrum analysis of different faults.

3.3. Wavelet Entropy of Different Transients

When using wavelet decomposition, the wavelet function and the number of decomposition levels need to be considered. The fourth Daubechies wavelet "db4" is used in the wavelet decomposition in this paper [41]. The number of wavelet decomposition levels should be less than nine levels [40]. Combined with the spectrum displayed in Figure 11, the number of decomposition levels is selected as eight. The wavelet entropy of different faults is shown in Figure 12.

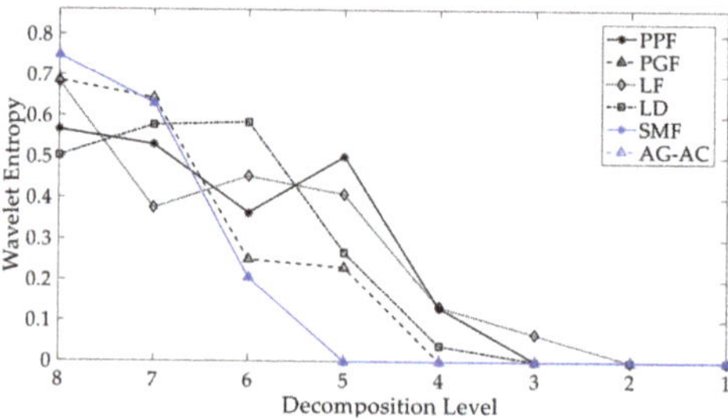

Figure 12. Wavelet entropies of different transients.

The wavelet entropy of each wavelet decomposition level indicates the degree of confusion of the detailed coefficients of the corresponding frequency band. From Figures 11 and 12, it can be seen that the amplitude and variation of the spectrum from 3125 Hz (200 kHz/2^{5+1}) to 50 kHz (200 kHz/2^2) of SMF and AG-AC are small. The wavelet entropy of the 2nd to 5th levels are all 0. This means that after filtering by the boundary, from the fifth to the first level, the differences between the content of different frequencies are too small that the calculated entropy value is 0. Its energy is mainly concentrated in the low-frequency band. The amplitude and change in the frequency spectrum of PGF are close to zero from 6250 Hz $(200 \text{ kHz}/2^{4+1})$ to 50 kHz $(200 \text{ kHz}/2^2)$. The wavelet entropies of the 2nd to 4th levels are all 0. This means that its transient signal energy and changes are mainly concentrated in the mid-frequency and low-frequency parts. From 12,500 Hz $(200 \text{ kHz}/2^{3+1})$ to 50 kHz $(200 \text{ kHz}/2^2)$, the amplitude and change of the frequency spectrum of LD are also small. Its wavelet entropy of the 2nd to 3rd levels is 0. Thus, its transient signal energy and changes are concentrated in the mid-frequency and high-frequency parts. The frequency spectrum of LF is close to zero from 25,000 Hz $(200 \text{ kHz}/2^{2+1})$ to 50,000 Hz $(200 \text{ kHz}/2^2)$. Its wavelet entropy of the 2nd is 0. Its transient signal energy and changes still exist in the high-frequency part. The wavelet entropies of PPF, from the 4th to 8th levels, are relatively large. This indicates that the content of each frequency band changes uniformly.

Different transients have clear differences in wavelet entropy. Combined with the distribution of wavelet entropy, it is expected that the precise action of protection will be realized. Furthermore, the transient frequency spectrum is affected by many factors, and it is necessary to analyze the influence of different influencing factors on the wavelet entropy.

3.4. Effect of Distance

As transient signals are analyzed in this research, the effect of distance cannot be ignored. Longer transmission distance suggests greater attenuation which will lead to waveform distortion, as well as frequency spectrum variation. Reliable judgments are required for protection. To analyze the characterization of wavelet entropy for transient signals in different fault distances, the wavelet entropy of different fault distances is discussed.

Figure 13 shows the distribution of wavelet entropy when the internal fault occurs in different locations. The faults are located at 20, 100, and 180 km along the line. As shown in Figure 13a, at different fault distances, the wavelet entropies of the 2nd to 4th levels are 0. From the 6th to 8th level, the wavelet entropies have some fluctuations. In Figure 13b, from the 5th to 8th level, the wavelet entropies have some fluctuations, but they are larger than 0 and are different from external faults. In Figure 13c, the values of the 2nd to 3rd levels are 0, the entropy of each level changes insignificantly, and the overall trend is consistent. In Figure 13d, the values of the 3rd level are larger than zero and different from LDs. Fluctuations are mainly concentrated from levels 5 to 8. Although the wavelet entropy changes under some decomposition levels, the overall distributions of wavelet entropy under different propagation distances are similar.

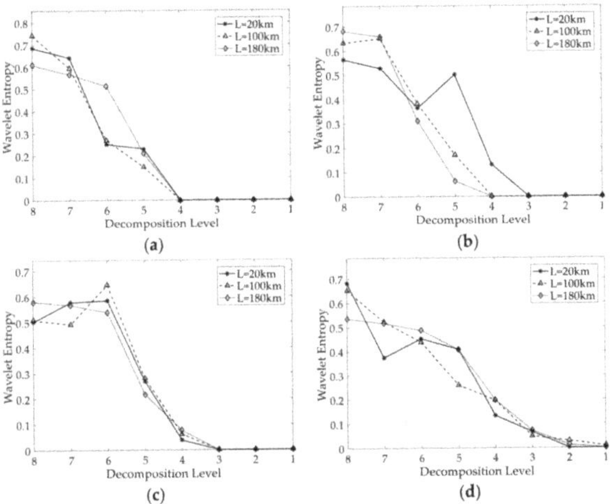

Figure 13. Effect of distance. (**a**) PGF; (**b**) PPF; (**c**) LD; (**d**) LF.

3.5. Effect of Ground Resistor

The ground resistor affects the sensitivity of the protection. An excessively large ground resistor may make the transient characteristics of the fault unobvious and cause the protection to fail. Figure 14 shows the distribution of wavelet entropy when the internal fault occurs with different ground resistors. The resistances of the ground resistor studied in this section are 0.001, 1, and 50 Ω. They are located at 100 km of the overhead lines. Figure 14a shows entropies of PGF; the values of the 2nd to 4th levels are 0. The trend of wavelet entropy changes is predominantly the same, and excessive resistor has little effect on the entropy. Figure 14b shows entropies of PPF; the values of the 5th level are larger than 0, are different from the external faults, and all values fluctuates slightly. Figure 14c shows entropies of LF; the wavelet entropy of each level fluctuates slightly, but the trend is the same. The values of the 3rd level are larger than 0 as well, and different from the LDs. Although the wavelet entropy has small numerical changes in some levels, the value has a small difference, and its overall distribution is the same.

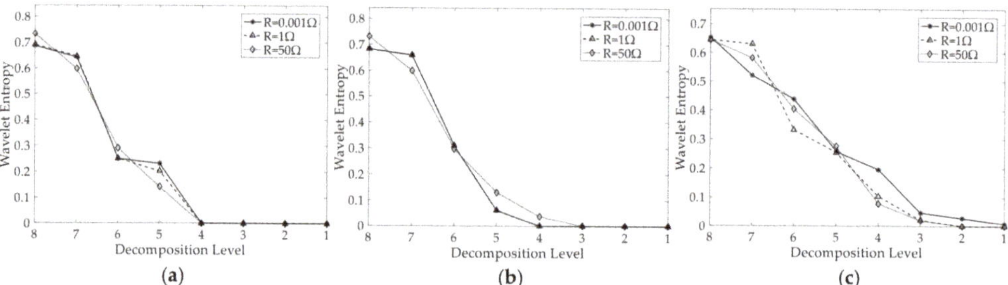

Figure 14. Effect of ground resistor. (**a**) PGF; (**b**) PPF; (**c**) LF.

From the analysis, it can be seen that wavelet entropy is not affected by distance and a certain range of ground resistor.

As discussed in this section, different types of faults have different distributions of wavelet entropies. The 5th level wavelet entropies of the external faults such as SMFs and AG-ACs are 0, while it is not zero for internal faults. The 4th level wavelet entropy of PGFs is 0, which is different from that of LDs and LFs. The 3rd level wavelet entropy of LDs is 0. It is different from the 3rd level wavelet entropy of LFs. Therefore, through the comparison of wavelet entropies in certain decomposition levels, it is possible to identify internal faults, external faults, and the fault types.

4. Methods of Transmission Line Protection

4.1. Starting Criterion

The starting element contains one transient criterion and one substitute criterion. The specific starting criterion is

$$|\Delta i| > 0.1 I_n \text{ or } i > 1.1 I_n \tag{5}$$

where Δi is the change in positive current. It is calculated by subtracting the value of the instantaneous current 1 ms before. i is the instantaneous current. I_n is the current rating.

The purpose of setting the starting element is to enhance the reliability of the protection.

4.2. Internal and External Fault Criterion

According to the wavelet entropy distribution in Figure 13, the high-frequency energy of external transients is filtered and attenuated due to the existence of boundaries or discontinuous boundaries, such as current-limiting reactors, SM capacitances, bridge arm reactors, etc. According to the analysis in Section 3.3, the value of the 5th wavelet entropy can be used as a criterion for distinguishing internal and external faults. The fault criterion is set as

$$W5 = 0, \tag{6}$$

where W5 represents the wavelet entropy value of the 5th level of the current line mode component during the fault. Therefore, if W5 = 0, it is considered as an external fault. Otherwise, it is determined as an internal fault.

4.3. PPF Criterion

Through the analysis in Section 2.2, in the initial stage of a PPF, there is a phenomenon of SM discharge. The fault current contains a large DC component. Through spectrum analysis, the fault current contains different frequency bands components with a large amplitude. The amplitude of the 0 Hz component can clearly distinguish the PPF. The fault criterion is set as

$$A0 > 2.7 \text{ kA}, \tag{7}$$

where A0 is the amplitude of the 0 Hz component. Through a large number of simulation experiments, the setting value is 2.7 kA. Therefore, if A0 > 2.7 kA, it is considered as a PPF. Otherwise, it is considered as another fault.

4.4. Recognition Method for PGF, LD, and LF

Combined with the analysis in Section 3, the spectrum distributions of PGF, LD, and LF have clear differences from 6250 Hz (200 kHz/2^{4+1}) to 25,000 Hz (200 kHz/2^{2+1}). Therefore, the wavelet entropy values of the 3rd to 4th levels have distinguished differences accordingly. This paper takes the value of the 4th level of wavelet entropy as the criterion of PGF. The fault criterion is set as

$$W4 = 0, \tag{8}$$

where W4 represents the wavelet entropy value of the 4th level of the current line mode component during the fault. Therefore, if W4 = 0, it is considered as a PGF. Otherwise, it is considered as a lightning strike.

According to the previous analysis, this paper adopts the value of the 3rd level of wavelet entropy as the criterion for the LD. The fault criterion is set as,

$$W3 = 0, \tag{9}$$

where W3 represents the wavelet entropy value of the 3rd level of the current line mode component during the fault. If W3 = 0, it is considered as an LD. Otherwise, it is considered as an LF.

4.5. Flow Chart of Protection Plan

Combined with the above analysis, the protection scheme is designed as shown in Figure 15. Firstly, the current value is detected to determine whether the protection system is activated or not. Then, the wavelet entropy of the line mode component is calculated and used to determine its fault zone based on the internal and external fault criterion. If it is an internal fault, the fault will be judged according to the PPF criterion, PGF criterion, and LD criterion in turn. Finally, the fault type is given.

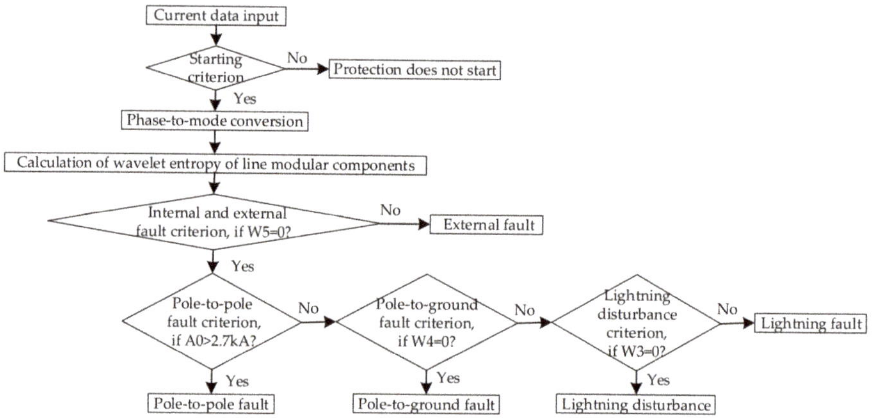

Figure 15. Flowchart of the protection plan.

5. Simulation and Discussion

A two-terminal MMC-HVDC system is modeled on the PSCAD platform, and the topology is as shown in Figure 1. The overhead transmission line model in PSCAD is applied. The line model is shown in Appendix A. The system parameters are show in Table 1.

Table 1. The parameters of the system.

Transmission line length	200 km	Transmission capacity	1000 MW
AC voltage	220 kV	DC bus voltage	±320 kV
Inductance of the smoothing reactor	50 mH	Inductance of the arm reactor	200 mH
Number of SMs of each bridge arm	12	SM capacitance	360 μF

This section simulates PPF, PGF, LF, FD, SMF, and AG-AC, with different parameters. The locations of those faults are demonstrated in Figure 1. Among them, the ground resistors are from 0.001 to 50 Ω. The distance to the fault location is from 0 to 200 km. The number of short-circuit modules of the SMF varies from 1 to 12. When the LF occurs, the lightning current amplitude ranges from 25 to 50 kA. When the LD occurs, the lightning current amplitude ranges from 5 to 30 kA. The details of simulation parameters are listed in Table 2.

Table 2. Experiment in different scenarios.

Type	Ground Resistor (Ω)/No. of Fault SMs	Fault Distance (km)	Amplitude of Lightning Strikes (kA)	No. of Faults
PPF	0.001, 0.1, 1, 10, 50	10, 20, 30, 40, 50, 60, 70, 80, 90, 100, 110, 120, 130, 140, 150, 160, 170, 180, 190, 200	N/A	100
PGF	0.001, 0.1, 1, 10, 50	10, 20, 30, 40, 50, 60, 70, 80, 90, 100, 110, 120, 130, 140, 150, 160, 170, 180, 190, 200	N/A	100
LD	N/A	10, 20, 30, 40, 50, 60, 70, 80, 90, 100, 110, 120, 130, 140, 150, 160, 170, 180, 190, 200	5, 10, 20, 30	80

Table 2. Cont.

Type	Ground Resistor (Ω)/No. of Fault SMs	Fault Distance (km)	Amplitude of Lightning Strikes (kA)	No. of Faults
LF	0.001, 0.1, 1, 10, 50	10, 20, 30, 40, 50, 60, 70, 80, 90, 100, 110, 120, 130, 140, 150, 160, 170, 180, 190, 200	25, 30, 40, 50	400
SMF	1, 2, 3, 4, 5, 6, 7, 8, 9, 10, 11, 12 (Six bridge arms)	N/A	N/A	72
AG-AC	0.001, 0.1, 1, 10, 50 (Three phases)	N/A	N/A	15

5.1. Protection Operation Results of Different Faults

Using the simulated MMC-based transmission model and simulated faults with parameters in Table 2, the protection operations of the proposed method are tested with different types of faults. The protection operation results are shown in Table 3.

Table 3. Results for different fault types with wavelet entropy.

Type	W5	Amplitude (0 Hz)	W4	W3	Result	Accuracy
PPF	>0	>2.7 kA	-	-	100 (PPF)	100%
PGF	>0	<2.7 kA	=0	-	100 (PGF)	100%
LD	>0	<2.7 kA	>0	=0	80 (LD)	100%
LF	>0	<2.7 kA	>0	>0	400 (LF)	100%
SMF	=0	N/A	N/A	N/A	72 (SMF)	100%
AG-AC	=0	N/A	N/A	N/A	15 (AG-AC)	100%

As shown in Table 3, the proposed protection method can effectively discriminate internal and external faults, and it can also recognize the internal transient type. No misoperations are generated by the wavelet entropy-based method. This method is effective in protecting non-faulted parts and finding faulted ones.

5.2. Effect of Number of Faulted SMs

The number of faulted SMs will affect the value and change of the short circuit current in DC transmission lines. It is necessary to study the effect of the number of fault SMs. Here, the performance of the proposed protection method is discussed as the number of fault SMs ranges from 1 to 12. The 5th level wavelet entropy W5 of fault current I_{dc} is always 0, which is obviously different from the case of an internal fault. It meets the internal and external fault criterion and is judged to be an external fault. As shown in Figure 16, the accuracy of the protection actions of different kinds of faults are all 100%, no matter how many SMs are faulted. Here, AP, AN, BP, BN, CP, and CN represent the upper (positive) and lower (negative) bridge arms of the ABC three-phase in the MMC converter, respectively. The change in the number of fault SMs will not affect the action of the protection criterion.

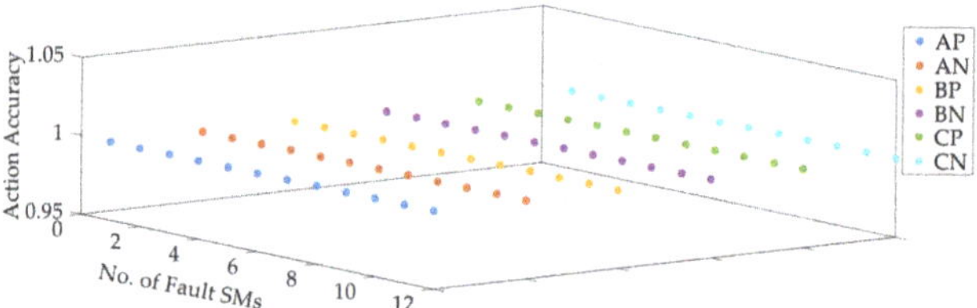

Figure 16. The influence of number of faulted SMs on the action accuracy.

5.3. Effect of Ground Resistor

A different ground resistor has an effect on the magnitude of transient signals, which will affect the accuracy of the protection actions. Here, the performance of the proposed method in dealing with PPF, PGF, LF, and AG-AC is discussed. Figure 17 shows the influence of the ground resistor on the action accuracy. As demonstrated by Figure 17, when the ground resistor is smaller than 50 Ω, the proposed method is immune to the change in ground resistor. However, with the increase in ground resistor, the fluctuation of high-frequency components of the transient currents becomes less obvious, which mean less disorder in high-frequency bands, and decreases. For internal faults, the values of W4 and W3 will decrease, and misoperations of protection will be found. However, for external faults, such as AG-AC, the criterion W5 = 0 is not affected. The judgment of external faults is still effective.

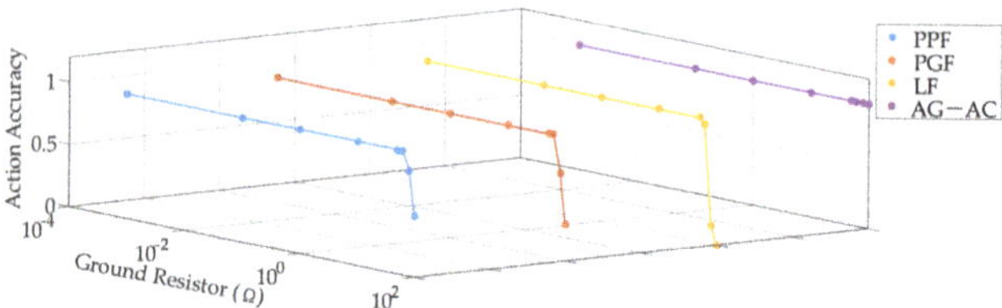

Figure 17. The influence of ground resistor on the action accuracy.

5.4. Effect of Fault Distance

The fault distance will affect the sensitivity of the protection scheme. It is necessary to study the effect of different fault distances for PPF, PGF, LD, and LF in this work. The action accuracy of the protection scheme when different faults occur at different positions is shown in Figure 18. When internal faults occur at different distances, the inequality W5 > 0 is always true. Under different internal faults, the values of A0, W4, and W3 are respectively in accordance with different fault type criterions. The fault types can be distinguished effectively without being affected by the fault distance. The results show the proposed wavelet entropy-based method is immune to the variation of fault distance.

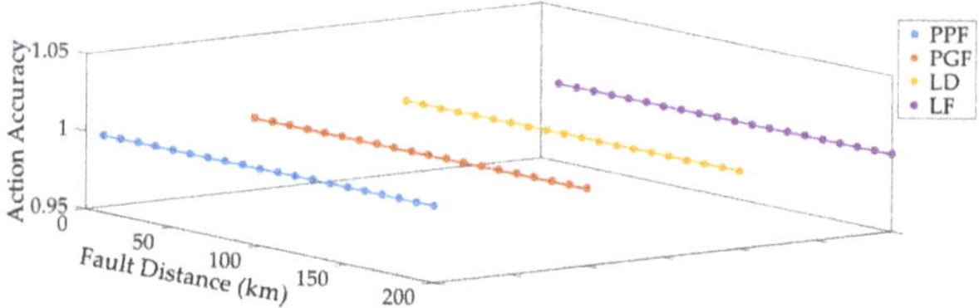

Figure 18. The influence of fault distance on the action accuracy.

5.5. Effect of Signal-To-Noise Ratio (SNR)

Noise interference is usually encountered in the practical environment. The influence of Gaussian white noise with different power on the protection is studied. The accuracy of the actions of the protection system under the influence of SNR is shown in Figure 19. As the Gaussian white noise will pollute the high frequency components of transients and the 5th level wavelet entropy will increase accordingly, the external faults will be judged to be internal ones. The noise tolerance of the internal and external fault criterion is only 60 dB. For internal faults recognition, the noise tolerance of the PPF criterion is 30 dB. It has good noise tolerance. The noise tolerance of the PGF criterion is 45 dB. When the noise reaches 30 dB, it will have a misjudgment rate up to 25%. The noise tolerance of the LD criterion and LF criterion is 30 dB. This means the protection method can effectively distinguish faults from disturbances even when the signal is seriously polluted.

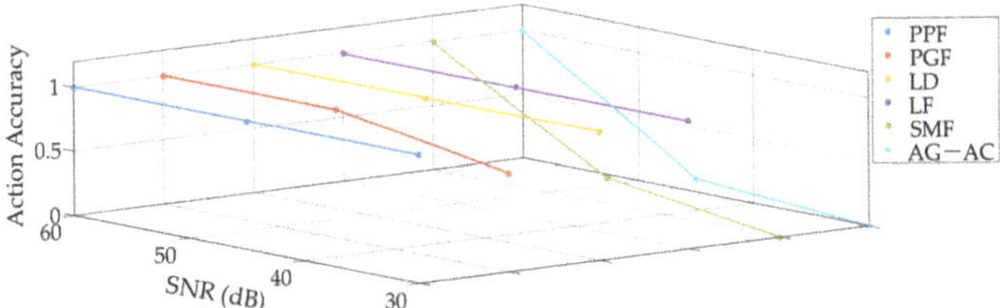

Figure 19. The influence of signal-to-noise (SNR) on the action accuracy.

From the above analysis, the proposed protection scheme is proved to be feasible in the absence of noise. The protection method can accurately distinguish the transient type under different scenarios, such as different ground resistors, different fault distances, different numbers of faulted SMs, and different SNRs. These results demonstrate the proposed wavelet entropy-based method can distinguish faults from interferences with high accuracy and reduce the misjudgments of protections.

6. Conclusions

A wavelet entropy-based protection method is proposed for symmetrical monopole MMC-HVDC overhead lines in this paper. Firstly, the fault characteristics of PPF, PGF, LD, LF, SMF, and AG-AC are analyzed in detail. Then, the definition of wavelet entropy is introduced and the wavelet entropy of different waveforms is analyzed. According to the different wavelet entropy values, the corresponding protection method is designed. Finally, theoretical analysis, simulation results generated by PSCAD/EMTDC, and discussions on the influence of different factors on the proposed protection method demonstrate that this method can not only accurately identify internal faults and external faults, but also effectively distinguish PPF, PGF, LD, and LF. At the same time, the method has strong adaptability and good noise tolerance in different fault locations and different fault ground resistors.

For systems with similar topologies, where same transient post-fault procedure can be produced, the proposed protection method can be used directly. However, for some other HVDC systems, such as asymmetrical monopole transmission system or bipolar asymmetrical transmission systems, the transient characteristics are different. The settings or criterions of the proposed wavelet-entropy-based protection method should be adjusted according to the spectrum distributions of different transients.

The research in this article is not to deny the application of traditional protection methods in MMC-HVDC, but to provide a new idea and perspective. Future work can focus more on improving the noise tolerance of the protection or introduce artificial intelligence methods to classify the characteristics of different faults.

Author Contributions: W.H. wrote the paper; G.L. developed the method; M.C. performed the experiments; J.H. provided some suggestions to improve this paper; Z.L. helped improve the expression of this paper; Y.Z. helped improve the performance of the algorithm. All authors have read and agreed to the published version of the manuscript.

Funding: This research was funded by the National Key R&D Program of China, grant number 2018YFB0904600.

Institutional Review Board Statement: Not applicable.

Informed Consent Statement: Not applicable.

Data Availability Statement: The data presented in this study are available on request from the corresponding author. The data are not publicly available due to privacy.

Acknowledgments: This research is funded by the National Key R&D Program of China (No. 2018YFB0904600).

Conflicts of Interest: The authors declare no conflict of interest.

Appendix A

The line model of overhead lines is shown in Figure A1. The tower data of the MMC-HVDC transmission line are shown in Figure A2.

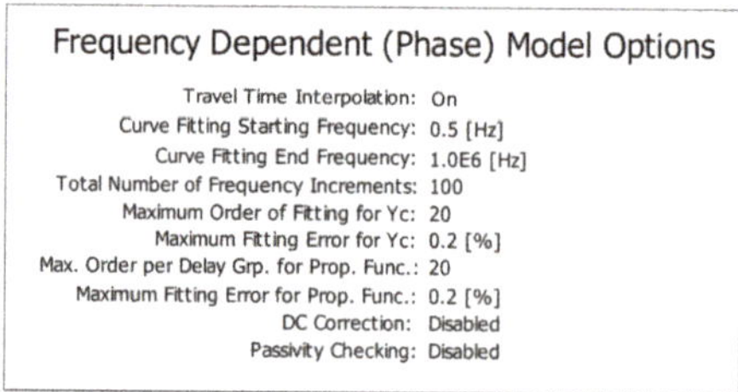

Figure A1. Line model of overhead lines.

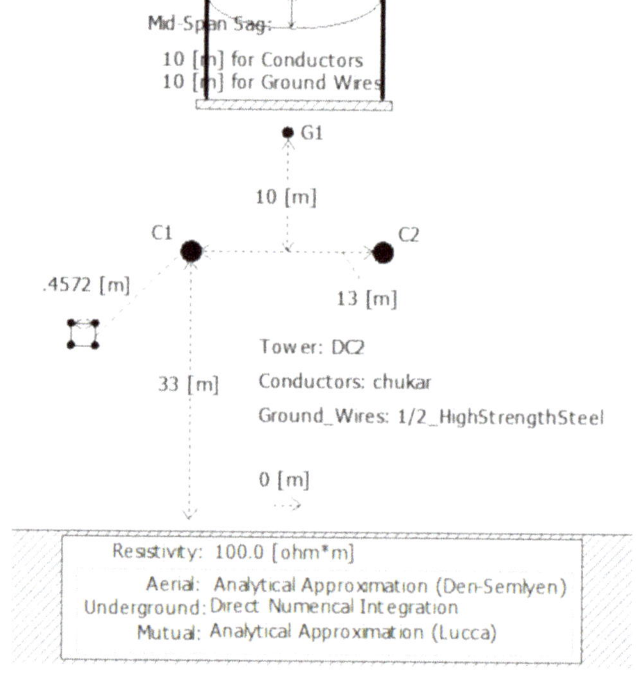

Figure A2. Tower data of the MMC-HVDC transmission line.

References

1. Kamran, S.; Lennart, H.; Hans-Peter, N.; Staffan, N.; Remus, T. Control and protection of MMC-HVDC under AC and DC network fault contingencies. In *Design, Control, and Application of Modular Multilevel Converters for HVDC Transmission Systems*; IEEE: Piscataway, NJ, USA, 2016; pp. 318–335.
2. Ning, L.; Zheng, X.; Tai, N.; Huang, W.; Chen, J.; Wu, Z. A novel pilot protection scheme for MMC-HVDC transmission lines. In Proceedings of the 2017 IEEE Energy Conversion Congress and Exposition (ECCE), Cincinnati, OH, USA, 1–5 October 2017; pp. 105–110.
3. Sneath, J.; Rajapakse, A.D. DC fault protection of a nine-terminal MMC HVDC grid. In Proceedings of the 11th IET International Conference on AC and DC Power Transmission, Birmingham, UK, 10–12 February 2015; pp. 1–8.
4. Cai, L.; Karaagac, U.; Mahseredjian, J. Simulation of startup sequence of an offshore wind farm With MMC-HVDC grid connection. *IEEE Trans. Power Deliv.* **2017**, *32*, 638–646. [CrossRef]
5. Karaagac, U.; Mahseredjian, J.; Cai, L.; Saad, H. Offshore wind farm modeling accuracy and efficiency in MMC-based multiterminal HVDC connection. *IEEE Trans. Power Deliv.* **2017**, *32*, 617–627. [CrossRef]
6. Bertilson, K.; Alishah, S.R.; Hosseini, H.S.; Babaei, E.; Aalami, M.; Ali, S.M.J.; Gharehpetian, B.G. A new generalized cascade multilevel converter topology and its improved modulation technique. *Int. J. Circ. Theor. Appl.* **2020**. [CrossRef]
7. Daryabak, M.; Filizadeh, S.; Jatskevich, J.; Davoudi, A.; Saeedifard, M.; Sood, V.K.; Martinez, J.A.; Aliprantis, D.; Cano, J.; Mehrizi-Sani, A. Modeling of LCC-HVDC systems using dynamic phasors. *IEEE Trans. Power Deliv.* **2014**, *29*, 1989–1998. [CrossRef]
8. Wu, J.; Li, H.; Wang, G.; Liang, Y. An improved traveling-wave protection scheme for LCC-HVDC transmission lines. *IEEE Trans. Power Deliv.* **2017**, *32*, 106–116. [CrossRef]
9. Belda, N.A.; Plet, C.A.; Smeets, R.P.P. Analysis of faults in multiterminal HVDC grid for definition of test requirements of HVDC circuit breakers. *IEEE Trans. Power Deliv.* **2018**, *33*, 403–411. [CrossRef]
10. Bucher, M.K.; Franck, C.M. Contribution of fault current sources in multiterminal HVDC cable networks. *IEEE Trans. Power Deliv.* **2013**, *28*, 1796–1803. [CrossRef]
11. Wen, W.; Huang, Y.; Sun, Y.; Wu, J.; Al-Dweikat, M.; Liu, W. Research on current commutation measures for hybrid DC circuit breakers. *IEEE Trans. Power Deliv.* **2016**, *31*, 1456–1463. [CrossRef]
12. Franck, C.M. HVDC circuit breakers: A review identifying future research needs. *IEEE Trans. Power Deliv.* **2011**, *26*, 998–1007. [CrossRef]
13. Kerf, D.K.; Srivastava, K.; Reza, M.; Bekaert, D.; Cole, S.; van Hertem, D.; Belmans, R. Wavelet-based protection strategy for DC faults in multi-terminal VSC HVDC systems. *IET Gener. Transm. Distrib.* **2011**, *5*, 496–503. [CrossRef]
14. Liu, X.; Osman, A.H.; Malik, O.P. Hybrid traveling wave/boundary protection for monopolar HVDC line. *IEEE Trans. Power Deliv.* **2009**, *24*, 569–578. [CrossRef]
15. Zhang, Y.; Tai, N.; Xu, B. Fault analysis and traveling-wave protection scheme for bipolar HVDC lines. *IEEE Trans. Power Deliv.* **2012**, *27*, 1583–1591. [CrossRef]
16. Suonan, J.; Gao, S.; Song, G.; Jiao, Z.; Kang, X. A novel fault-location method for HVDC transmission lines. *IEEE Trans. Power Deliv.* **2010**, *25*, 1203–1209. [CrossRef]
17. Ikhide, M.; Tennakoon, S.; Griffiths, A.; Subramanian, S.; Ha, H. Fault detection in Multi-Terminal Modular Multilevel Converter (MMC) based High Voltage DC (HVDC) transmission system. In Proceedings of the 2015 50th International Universities Power Engineering Conference (UPEC), Piscataway, NJ, USA, 1–4 September 2015; pp. 1–6.
18. Leterme, W.; Beerten, J.; Hertem, D.V. Nonunit protection of HVDC grids with inductive DC cable termination. *IEEE Trans. Power Deliv.* **2016**, *31*, 820–828. [CrossRef]
19. Tzelepis, D.; Dyśko, A.; Fusiek, G.; Nelson, J.; Niewczas, P.; Vozikis, D.; Orr, P.; Gordon, N.; Booth, D.C. Single-ended differential protection in MTDC networks using optical sensors. *IEEE Trans. Power Deliv.* **2017**, *32*, 1605–1615.
20. Samir, A.; Abu-Elanien, A.E.B.; Abdel-Khalik, A.S.; Massoud, A.M.; Ahmed, S. A directional protection technique for MTDC networks. In Proceedings of the 2015 4th International Conference on Electric Power and Energy Conversion Systems (EPECS), Sharjah, United Arab Emirates, 24–26 November 2015; pp. 1–6.
21. Song, G.; Hou, J.; Guo, B.; Chen, Z. Pilot protection of hybrid MMC DC grid based on active detection. *Prot. Control. Mod. Power Syst.* **2020**, *5*, 6. [CrossRef]
22. Li, Y.; Gong, Y.; Jiang, B. A novel traveling-wave-based directional protection scheme for MTDC grid with inductive DC terminal. *Electr. Power Syst. Res.* **2018**, *157*, 83–92. [CrossRef]
23. He, J.; Chen, K.; Li, M.; Luo, Y.; Liang, C.; Xu, Y. Review of protection and fault handling for a flexible DC grid. *Prot. Control. Mod. Power Syst.* **2020**, *5*, 15. [CrossRef]
24. Li, B.; He, J.; Li, Y.; Li, B. A review of the protection for the multi-terminal VSC-HVDC grid. *Prot. Control. Mod. Power Syst.* **2019**, *4*, 21. [CrossRef]
25. Gaouda, A.M.; El-Saadany, E.F.; Salama, M.M.A.; Sood, V.K.; Chikhani, A.Y. Monitoring HVDC systems using wavelet multi-resolution analysis. *IEEE Trans. Power Syst.* **2001**, *16*, 4–662. [CrossRef]
26. Luo, G.M.; Yao, C.Y.; Liu, Y.L.; Tan, Y.J.; He, J.H. Entropy SVM-based recognition of transient surges in HVDC transmissions. *Entropy Artic.* **2018**, *20*, 421. [CrossRef] [PubMed]

27. Chen, B.; Wang, J.; Zhao, H.; Principe, J.C. Insights into entropy as a measure of multivariate variability. *Entropy* **2016**, *18*, 196. [CrossRef]
28. Chen, B.; Zhu, Y.; Hu, J. Mean-square convergence analysis of ADALINE training with minimum error entropy criterion. *IEEE Trans. Neural Netw.* **2010**, *21*, 1168–1179. [CrossRef] [PubMed]
29. Ruiye, L.; Xin, S.; Zhimin, L. On the application of entropy in excitation control. In Proceedings of the 2004 International Conference on Power System Technology, PowerCon, Singapore, 21–24 November 2004; Volume 1, pp. 952–956.
30. Chen, J.; Li, G. Tsallis wavelet entropy and its application in power signal analysis. *Entropy* **2014**, *16*, 3009–3025. [CrossRef]
31. Luo, G.; Zhang, D.; Koh, Y.; Ng, K.; Leong, W. Time—Frequency entropy-based partial-discharge extraction for nonintrusive measurement. *IEEE Trans. Power Deliv.* **2012**, *27*, 1919–1927. [CrossRef]
32. Aggarwal, R.K.; Johns, A.T.; Bo, Z.Q. Non-unit protection technique for EHV transmission systems based on fault-generated noise, Part 2: Signal processing. *IEE Proc. Gener. Transm. Distrib.* **1994**, *141*, 141–147. [CrossRef]
33. Johns, A.T.; Martin, M.A.; Barker, A.; Walker, E.P.; Crossley, P.A. A new approach to E.H.V. direction comparisn protection using digital signal processing techniqles. *IEEE Trans. Power Deliv.* **1986**, *1*, 24–34. [CrossRef]
34. Qiang, G.; Xi, Y.; Ye, L. Circulating current suppressing and AC faults ride-through capability analysis of Zhoushan MMC-MTDC system. In Proceedings of the 2018 2nd IEEE Conference on Energy Internet and Energy System Integration (EI2), Beijing, China, 20–22 October 2018; pp. 1–6.
35. Abdalrahman, A.; Isabegovic, E. DolWin1-Challenges of connecting offshore wind farms. In Proceedings of the 2016 IEEE International Energy Conference (ENERGYCON), Piscataway, NJ, USA, 4–8 April 2016; pp. 1–10.
36. Ding, H.; Wu, Y.; Zhang, Y.; Ma, Y.; Kuffel, R. System stability analysis of Xiamen bipolar MMC-HVDC project. In Proceedings of the 12th IET International Conference on AC and DC Power Transmission (ACDC 2016), Beijing, China, 28–29 May 2016; pp. 1–6.
37. Hu, J.; Zhao, C.; Zhang, X.; Yang, X. Simulation study of the Zhoushan project as a three-terminal DC transmission system. In Proceedings of the 2012 IEEE Power and Energy Society General Meeting, San Diego, CA, USA, 22–26 July 2012; pp. 1–6.
38. You, M.; Zhang, B.H.; Cheng, L.Y.; Bo, Z.Q.; Klimek, A. Lightning model for HVDC transmission lines. In Proceedings of the 10th IET International Conference on Developments in Power System Protection (DPSP 2010). Managing the Change, Manchester, UK, 29 March–1 April 2010; pp. 1–5.
39. Luo, G.; Zhang, D.; Tseng, K.J.; He, J. Impulsive noise reduction for transient Earth voltage-based partial discharge using Wavelet-entropy. *IET Sci. Meas. Technol.* **2016**, *10*, 69–76.
40. Luo, G.M.; Lin, Q.Z.; Zhou, L.; He, J.H. Recognition of traveling surges in HVDC with wavelet entropy. *Entropy* **2017**, *19*, 184. [CrossRef]
41. Mallat, S.G. Wavelet basis. In *A Wavelet Tour of Signal Processing: The Sparse Way*; Elsevier: Amsterdam, The Netherlands; Academic Press: Boston, MA, USA, 2009; pp. 263–376.

Article

Numerical and Experimental Study of Lightning Stroke to BIPV Modules

Xiaoyan Bian [1], Yao Zhang [1], Qibin Zhou [1,2,*], Ting Cao [2] and Bengang Wei [3]

[1] College of Electrical Engineering, Shanghai University of Electric Power, Shanghai 200090, China; bianxy@shiep.edu.cn (X.B.); zhangyao8253@163.com (Y.Z.)

[2] School of Mechatronic Engineering and Automation, Shanghai University, Shanghai 200444, China; 15102348682@163.com

[3] State Grid Shanghai Municipal Electric Power Company, Shanghai 200122, China; wbgsj@126.com

* Correspondence: zhouqibin@shu.edu.cn; Tel.: +86-138-1779-0981

Abstract: Building Integrated Photovoltaic (BIPV) modules are a new type of photovoltaic (PV) modules that are widely used in distributed PV stations on the roof of buildings for power generation. Due to the high installation location, BIPV modules suffer from lightning hazard greatly. In order to evaluate the risk of lightning stroke and consequent damage to BIPV modules, the studies on the lightning attachment characteristics and the lightning energy withstand capability are conducted, respectively, based on numerical and experimental methods in this paper. In the study of lightning attachment characteristics, the numerical simulation results show that it is easier for the charges to concentrate on the upper edge of the BIPV metal frame. Therefore, the electric field strength at the upper edge is enhanced to emit upward leaders and attract the lightning downward leaders. The conclusion is verified through the long-gap discharge experiment in a high voltage lab. From the experimental study of multi-discharge in the lab, it is found that the lightning interception efficiency of the BIPV module is improved by 114% compared with the traditional PV modules. In the study of lightning energy withstand capability, a thermoelectric coupling model is established. With this model, the potential, current and temperature can be calculated in the multi-physical field numerical simulation. The results show that the maximum temperature of the metal frame increases by 16.07 °C when 100 kA lightning current flows through it and does not bring any damage to the PV modules. The numerical results have a good consistency with the experimental study results obtained from the 100 kA impulse current experiment in the lab.

Keywords: building integrated photovoltaic (BIPV); lightning attachment characteristics; lightning energy withstand capability; numerical and experimental analysis

Citation: Bian, X.; Zhang, Y.; Zhou, Q.; Cao, T.; Wei, B. Numerical and Experimental Study of Lightning Stroke to BIPV Modules. *Energies* **2021**, *14*, 748. https://doi.org/10.3390/en14030748

Academic Editor: Adam Dyśko

Received: 27 December 2020
Accepted: 27 January 2021
Published: 1 February 2021

Publisher's Note: MDPI stays neutral with regard to jurisdictional claims in published maps and institutional affiliations.

1. Introduction

As a new form of clear power generation, photovoltaic (PV) power generation has attracted more and more attention around the world. Currently, first and second-generation PV technologies are already included for building integration photovoltaic (BIPV) and building attached/applied photovoltaic (BAPV) application in the form of roof, window, wall and shading elements. In addition, third-generation PVs are under exploration [1–3]. With BIPV, it can not only generate power for the building consumers or even power companies, but also save cost and space when constructing the distributed PV stations on buildings. The application of BIPV modules is the development trend of green buildings. It also represents the future of urban and building energy development [4–6].

Due to the high installation location, BIPV modules suffer from lightning hazard greatly. Lightning is a natural phenomenon of a strong discharge, releasing tremendous energy. When lightning strikes BIPV, it will cause deformation and melting of metal framework and damage to PV modules [7,8]. In total, 5% to 10% of solar installations are damaged by direct lightning or lightning electromagnetic pulse every year [9].

To ensure PV systems safe and reliable, lightning protection design attracts more and more attention. At present, there is much research on direct lightning and lightning-induced overvoltage in PV power plants [10–13]. Y. G. Jia [14] and F. Y. Guo [15] pointed out the design of the green building and the key point of lightning protection. N. H. Zaini [16] simulated the lightning stroke of different waveforms and amplitudes on different parts of the photovoltaic system, and found that the transient current would appear at the nearest point to the lightning striking, while the transient voltage would appear on the AC side of the inverter. Jae-Young Cho [17] analyzed the damage of lightning overvoltage to PV array infrastructure, and considered the influence caused by different lightning striking locations and distances on PV array. C. Dechthummarong [18] studied the lightning withstand of insulation materials of field-aged PV modules, applied a pulse of 1.2/50 μs between impulse voltage generator (IVG) and PV modules, and realized the mathematical model of partial discharge in PV insulation gap. The experimental results showed that insulation defects caused by aging produce partial discharge between PV modules and aluminum frame. K. Tamura [19] conducted a lightning stroke test on a centralized PV system with an area of 75 m^2, a height of about 10 m and an installed capacity of 14 kW. The test showed that the lightning current attached the fabricated aluminum chassis and flowed into the ground, which verified the lightning energy withstand capability of centralized PV modules. K. M. Coetzer [20] conducted a lightning current test in a high voltage lab, and found that poor wiring between PV modules leads to the arising of the excessive induced current. Then, a better wiring method was proposed to reduce the amplitude of the induced current.

As an important part of the integrated roof, the BIPV modules face the threat of direct lightning strike. The configuration of BIPV is quite different from the traditional PV modules which are installed on the fixed PV brackets on the roof. However, the PV modules of BIPV are integrated with the metallic roof without any PV bracket. It changes the relative position between PV modules and surrounding metallic structures compared to the traditional PV modules. Therefore, the study of traditional PV modules is not available for BIPV directly. The study of lightning stroke to BIPV modules is very little addressed nowadays.

In order to protect BIPV modules against lightning damage, a study about the lightning stroke to BIPV modules is conducted with numerical and experimental methods in this paper. Since lightning effect to BIPV modules can be divided into two aspects, i.e., lightning attachment and lightning energy withstand, the study is also divided into two parts, the attachment characteristics in Section 2 and the energy withstand capability in Section 3 of this paper. The study conclusions of this paper will be guidance for the lightning protection of BIPV modules used in rapidly growing distributed PV stations. It would be expected that the combination of PV and building is one of the most important areas for future PV applications. It improves the lightning protection requirements of green buildings and PV systems.

2. Study of Lightning Attachment Characteristics to BIPV Modules

2.1. Numerical Simulation and Analysis

2.1.1. Modeling of Lightning Stroke to BIPV Modules

During a thunderstorm, the lightning downward leaders develop step by step from the cloud to the ground. As the lightning downward leaders approach to the ground, upward leaders will be generated when the electric field strength of the object on the ground reaches a certain level. Once the downward leader connects with the upward leader from a certain object such as BIPV modules on a building, the object will be struck by the lightning, as Figure 1 shows.

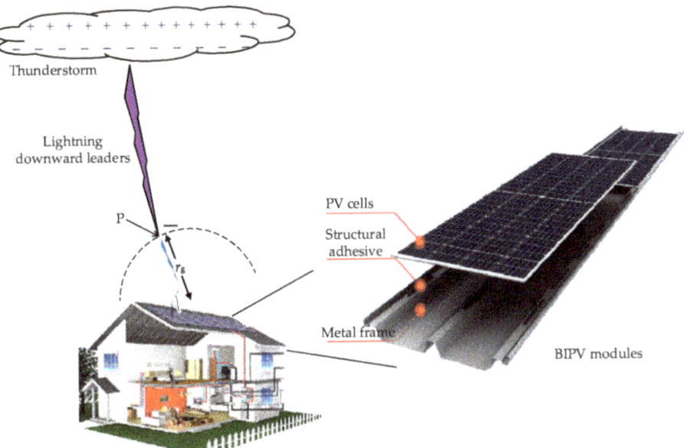

Figure 1. Schematic diagram of lightning stroke to building integration photovoltaic (BIPV) modules.

The connection is dependent on the distance between the downward leader and the upward leader reaching the breakdown threshold of the electric field. This threshold is defined as the lightning striking distance which can be interpreted as a function of lightning current, as Equations (1) and (2) shows.

$$r_c = KI_P^b \tag{1}$$

where r_c is the lightning striking distance, m; I_P is the amplitude of lightning current, kA; K and b are coefficients to account for different striking distances to a mast, a shield wire, or the ground plane [21].

According to buildings or transmission lines at different altitudes, scholars have carried out a lot of research on the three important parameters of the striking distance [22–27]. The lightning distance r_g should be corrected by Equation (2). The specific values of the lightning striking distance parameters from different literature are listed in Table 1.

$$r_g = K_g r_c \tag{2}$$

K_g is the correction factor to correct the striking distance r_c of the earth; r_g is the corrected lightning distance.

Table 1. Striking distance parameters from different literatures.

Literature	K	b	Kg
Golde [22]	3.30	0.78	1
Young [23]	27.00	0.32	1 (h < 18); 444/(462-h)(h > 18)
Love [24]	10.00	0.65	1
Whitehead [25]	9.4	0.67	1
Anderson [26]	8.00	0.65	0.64 (EHV lines); 0.8 (UHV lines); 1 (other lines)
IEEE std 2012 [27]	8	0.65	1

According to the striking distance parameters given in IEEE Std 998-2012 [27], where $K = 8$, $b = 0.65$, $K_g = 1$. Because 50% of current amplitude $I_P = 30$ kA [28], the lightning striking distance r_g is calculated to be 73 m.

The downward leader is simplified as a rod electrode when studying the lightning striking position. The electrode voltage is taken as 50% breakdown voltage of negative lightning discharge of the rod–rod gap in air. According to Equation (3), the electrode voltage is set as 43.9 MV.

$$U_{50\%} = 110 + 6d \tag{3}$$

where $U_{50\%}$ is 50% breakdown voltage of lightning impulse, kV. d is the distance between poles, cm.

In the FEA software, an electrostatic field model of PV modules is established under lightning downward leader. As shown in Figure 2, the downward leader head is equivalent to a rod electrode, and the relative spatial distance between the rod electrode and the PV module is r_g. The PV module consists of an inner solar cells and an outer metal frame as Figure 1 shows. The material of metal frame is set to aluminum. The material of solar cells is set to silicon. Moreover, the metal frame is set to zero potential as grounding. The lightning interception position of different PV modules can be determined by evaluating the distribution of static electric field strength on the PV modules. The position with the maximum field strength is prone to be struck with maximum possibility.

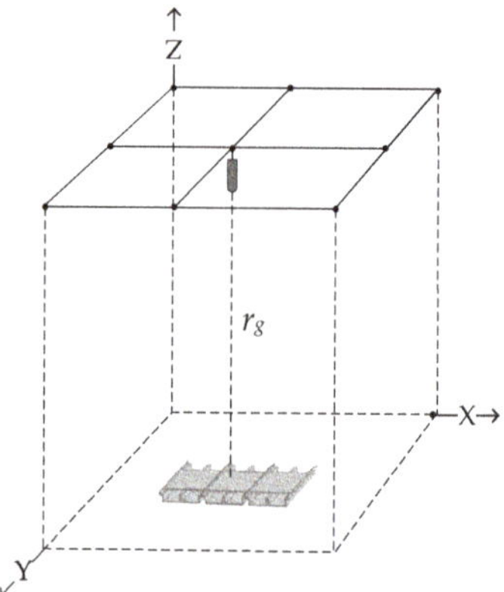

Figure 2. Simulation model of photovoltaic (PV) module under electrostatic field.

2.1.2. Electrostatic Field Theory

The electric field, due to a given charged lightning stepped leader, can be calculated using Equations (4)–(7) of electrostatics.

$$\nabla \times \boldsymbol{E} = 0 \tag{4}$$

$$\nabla \bullet \boldsymbol{D} = \rho \tag{5}$$

$$\boldsymbol{E} = -\nabla \varphi \tag{6}$$

$$\nabla^2 \varphi = -\frac{\rho}{\varepsilon} \tag{7}$$

where $\boldsymbol{E}$ is the electric field tensor; $\boldsymbol{D}$ is the electric displacement vector; ρ is the charge density; φ is the electric potential; and ε is the permittivity of the free space.

It should be noted that the lightning distance r_g determines the relative spatial position between the lightning downward leader and BIPV modules in the computational domain, and also determines the locations of the zero-electric potential boundary conditions that are associated with the metal frame.

2.1.3. Simulation Results and Analysis

The FEA software is used to solve the governing Equations (4)–(7). As shown in Figure 3a, the sharp corners of metal frame are easy to accumulate charges. The maximum surface charge density is 140 $\mu c/m^2$. The minimum surface charge density is 6.89 $\mu c/m^2$. As shown in Figure 3b, charges are more likely to accumulate on the border of BIPV modules, with the maximum surface charge density of 194 $\mu c/m^2$ and the minimum surface charge density of 0.11 $\mu c/m^2$. In the electrostatic field formed by the lightning downward leader head, the space charge accumulates on the surface of the ground object leading to the electric field. The upper edge of metal frame is easier to gather charges which can enhance electric field intensity.

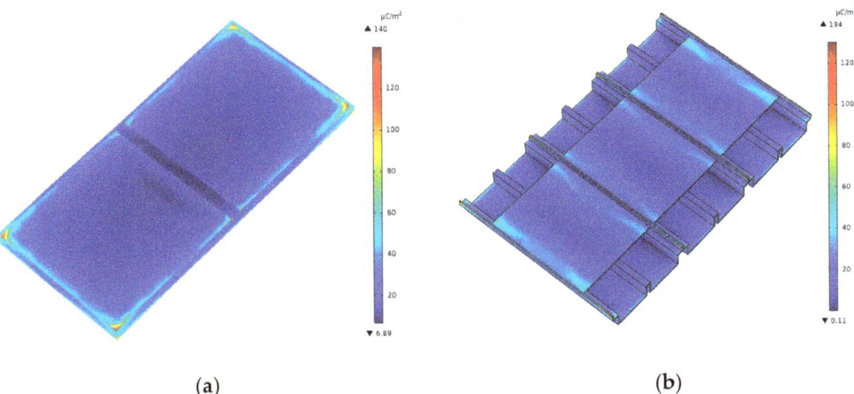

(a) (b)

Figure 3. Surface charge density of two PV modules (**a**) framed double-glass PV module; (**b**) BIPV module unit combination.

Through numerical calculation, the surface electric field strength distribution of the two PV modules is shown in Figure 4. It can be seen that the highest electric field strength of framed double-glass PV module is distributed at the four corners of the metal frame. The highest field strength of BIPV module is distributed on the upper edge of metal frame. In addition, the electric field strength on the two metal frames is much higher than that of the internal solar cells.

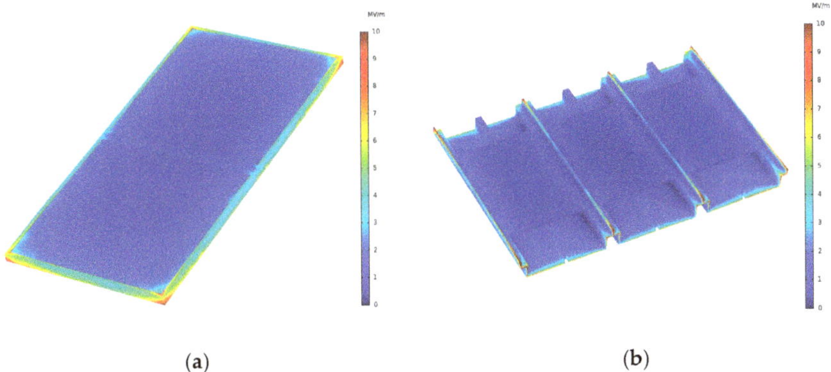

(a) (b)

Figure 4. Surface electric field strength of different PV modules (**a**) framed double-glass PV module; (**b**) BIPV module unit combination.

Take different paths along the arrow direction on the metal frame, defined as a1–a4 and b1–b4. The measured electric field strength along different paths is shown in Figure 5. The maximum electric field strength is distributed at both ends of the metal frame. Those on the surfaces of framed double-glass PV module and BIPV module is 9.5 MV/m and 22.1 MV/m respectively. The electric field strength is increased by 132%. The locations of the PV modules surface with higher electric field strength are considered to have higher possibility of emitting answering leaders and intercepting the lightning stepped leader.

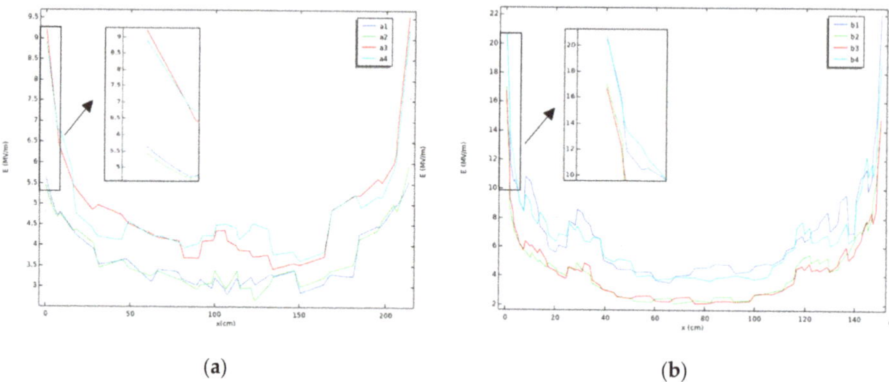

(a)   (b)

Figure 5. Distribution of surface electric field strength (**a**) framed double-glass PV module; (**b**) BIPV module unit combination.

2.2. Experimental Study

2.2.1. Sample Preparation

In order to compare the different lightning protection performance of various PV modules and verify the accuracy of the simulation, three experimental samples were selected according to the models in the numerical simulation in Section 2.1. As shown in Figure 6a, a 213 cm × 105 cm framed double-glass PV module was selected as a counterpart piece and defined as Sample A, which was assembled by the metal frame and solar cells. At the same time, a 222 cm × 150 cm BIPV module unit combination was defined as Sample B, as shown in Figure 6b. In the lightning current withstand capability test in Section 3.2, the BIPV module unit with metal frame of 150 cm × 72.5 cm was selected for the test, which is defined as Sample C, as shown in Figure 6c.

(a)   (b)

Figure 6. *Cont.*

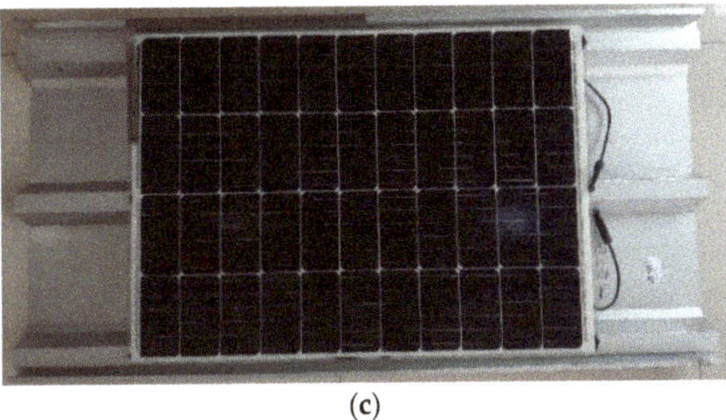

(c)

Figure 6. Experimental samples (**a**) framed double-glass PV module; (**b**) BIPV module unit combination; (**c**) BIPV module unit.

2.2.2. Experimental Method

In order to simulate the lightning stroke of PV modules under thunderstorm clouds, an experimental platform was set up in the high voltage laboratory. The long gap discharge of PV module by SJTU-3000 impulse voltage generator (IVG) was used to simulate the spatial potential and electric field distribution of PV modules during the scene of lightning downward leader descending process.

The schematic diagram of the experimental platform is shown in Figure 7a. Place Sample A and Sample B horizontally on the ground. Keep the metal frame well grounded. The rod electrode connected with SJTU-3000 IVG was located directly above the center of PV modules, with a vertical distance of 2 m. The layout of the experimental platform is shown in Figure 7b. As more than 90% of lightning in nature is negative, SJTU-3000 IVG was used to apply negative lightning impulse voltage with 1.2/50 μs waveform for 20 times. In order to prevent the subsequent experiment from being affected by the breakdown of the solar cells, the solar cells surface of the experimental sample was covered with 5 mm thick glass. Moreover, the insulation blanket was used to cover the sharp corners to prevent lightning striking at these corners because these sharp corners do not exit when BIPV is installed on the building roof. Therefore, the influence of glass and insulating blanket on this test can be ignored. After each discharge, the location of lightning striking attachment point was recorded to calculate the lightning striking attachment probability of the samples, with the lightning protection effect of different PV modules evaluated.

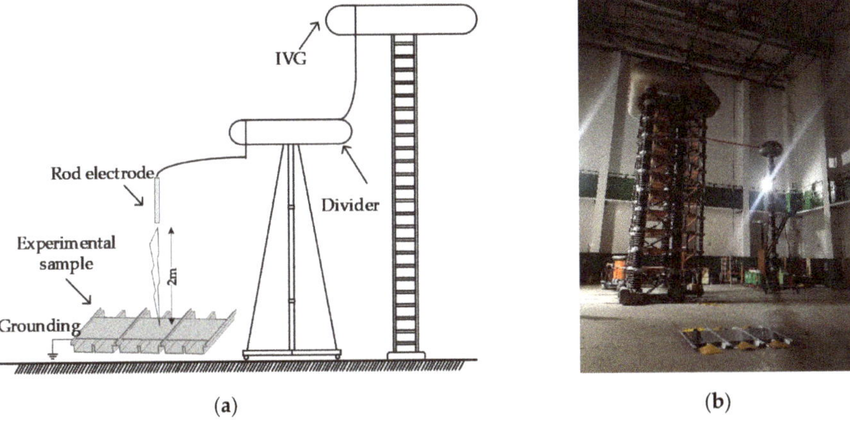

(**a**)   (**b**)

Figure 7. Lightning attachment experiment (**a**) Schematic diagram of experimental platform; (**b**) Layout of experimental setup.

2.2.3. Experimental Results and Analysis

According to the experiment method in Section 2.2.2, the long gap arc was attached to the metal frame of Sample A seven times. The arc was attached to the solar cells and flashed over the surface to the metal frame 13 times. Consequently, the interception efficiency of the metal frame was 35%. Figure 8a,b show the successful and failed interception of the discharge arc of Sample A. Under the same experimental conditions, it was noted that the arc attached to the metal frame of Sample B for 15 times. The other five discharge arcs flashed over the surface of solar cells. Then the interception efficiency is 75%. Figure 8c,d show the successful and failed interception of the discharge arc of Sample B. Compared with the electric field distribution and the experimental results, the location of lightning attachment point is very close to simulation results in Section 2.1.3, which can be used to check the effectiveness of the simulation method.

Figure 8. Lightning connection of different PV modules (**a**) Lightning connection of metal frame of Sample A; (**b**) Surface flashover of Sample A; (**c**) Lightning connection of metal frame of Sample B; (**d**) Surface flashover of Sample B.

It is noted that under the impulse 1.2/50 µs standard lightning voltage, the metal frame of Sample B was easier to attract the discharge arc generated by the rod electrode than the metal frame of Sample A. The interception efficiency is increased by 114%. That is to say, the metal frame of the BIPV module has better lightning interception performance, which can protect the solar cells from lightning damage more effectively.

3. Study of Lightning Energy Withstand Capability to BIPV Modules

3.1. Numerical Simulation and Analysis

3.1.1. Thermoelectric Coupling Theory

In the process of flowing through BIPV modules, lightning current converted to Joule heat acts on BIPV modules. Studying the temperature field distribution plays a crucial role in the lightning energy withstand capability of BIPV modules. In this section, the thermoelectric coupling analysis model of BIPV modules is established. The solution of the model is based on the law of charge conservation and the law of conservation of energy.

According to the law of charge conservation, the governing equation of electric field in conductor is as follows,

$$\int_{S} \boldsymbol{J} \bullet \mathbf{n} dS = \int_{v} r_c dV \tag{8}$$

where S is cross sectional area of unit; V is unit volume; $\mathbf{n}$ is normal vector; $\boldsymbol{J}$ is current density; and r_c is charge density per unit volume.

Based on Ohm's law, current density can be defined as Equations (9),

$$\boldsymbol{J} = \sigma \bullet \boldsymbol{E} = -\sigma \bullet \nabla \Phi \tag{9}$$

where σ is the conductivity; $\boldsymbol{E}$ is the electric field strength; and Φ is the electrical potential.

The governing equation in the finite element is obtained as Equations (10),

$$\int \nabla \delta \Phi \bullet (\sigma \bullet \nabla \Phi) dV = \int_{V} \delta \Phi r_c dV + \int_{S} \delta \Phi \bar{J} dS \tag{10}$$

According to Joule's law, the current flowing through the conductor will be converted into Joule heat, as shown in the following relationship.

$$P_{ec} = \boldsymbol{J} \bullet \boldsymbol{E} = (\sigma \bullet \boldsymbol{E}) \bullet \boldsymbol{E} \tag{11}$$

$$r = \eta_v P_{ec} \tag{12}$$

where P_{ec} is the power loss density; η_v is the energy conversion factor; and r is the thermal density.

The governing equation of heat conduction is obtained as Equations (13),

$$\int_{v} \rho C_v \frac{\partial \theta}{\partial t} \partial \theta dV + \int_{v} \nabla \partial \theta \bullet (K \bullet \nabla \theta) dV = \int_{v} \delta \theta r dV + \int_{s} \delta \theta q dS \tag{13}$$

where θ is the temperature; K is the solid thermal conductivity; ρ is the solid density; C_v is the specific heat capacity; and q is the heat flux density.

Equations (10) and (13) describe the governing equations of thermoelectric coupling problem. When studying lightning current withstand capability of BIPV modules, the potential, current and temperature can be calculated in the multi-physical field numerical simulation.

3.1.2. Modeling of Lightning Current

The heat source was given as a large current in the multi-physical field numerical simulation model. The lightning current used in lightning transient calculation can be divided into three categories: double exponential model, Heidler model and pulse function model [29]. In 1941, Bruce and Golde put forward the double exponential function model of lightning current [30]. The concise mathematical expression can describe the typical lightning current waveform, which has been widely used in the field of lightning protection. The expression form is expressed in Equation (14).

$$i(t) = \frac{I_0}{\eta} \left(e^{-\alpha t} - e^{-\beta t} \right) \tag{14}$$

In which I_0 is the peak current; η indicates the peak correction coefficient; $\alpha = 1/T_1$, $\beta = 1/T_2$, T_1 is the wave head time and T_2 is the wave tail time.

In the simulation, the mathematical model of lightning current adopts double exponential model, the amplitude of lightning current is 10–100 kA, the peak correction coefficient is 0.998, the current waveform is 10/350 µs standard lightning wave. According to the results of Section 2.2.3, the maximum electric field strength on the surface of PV module is set as the lightning injection point.

3.1.3. Simulation Result and Analysis

The methodology of Section 3.1.1 can be used to verify the transient state of current distribution and lightning energy withstand capability of BIPV module unit. Figure 9 shows the transient distribution of lightning current at six times. Lightning current flows from attachment point of the metal frame to the grounding point. From 0 µs to 14 µs, the current density instantly increases and reaches a peak value of 4.16×10^9 A/m^2. At the half peak time of 350 µs, the current density gradually decreases to 2.06×10^9 A/m^2, and then tends to 0. Because the simulated lightning current waveform is 10/350 µs with a short rising edge, the current density reaches the maximum in a short time. There is no current passing through the solar cells in the whole process, so it is noted that the metal frame can effectively protect the solar cells.

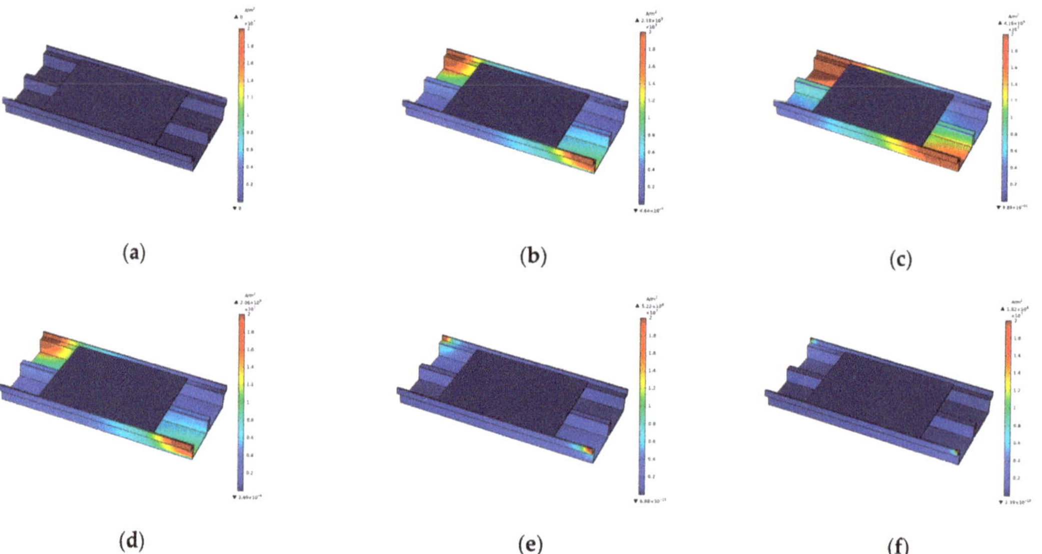

Figure 9. Current transient distribution of BIPV at different times (**a**) 0 µs; (**b**) 2 µs; (**c**) 14 µs; (**d**) 350 µs; (**e**) 1000 µs; (**f**) 1500 µs.

In addition, the temperature rise effect of BIPV during lightning striking is shown in Figure 10. After 1200 µs, the amplitude of lightning current is reduced to 10% I_{imp}. The resistivity of the metal frame is very low and the whole process is transient. Therefore, the temperature rise tends to be stable after 1200 µs. The maximum temperature rise only increases from 0.17 °C to 16.07 °C with increasing lightning current from 10 kA to 100 kA. This temperature rise is easy for the metal frame to withstand without causing any damage.

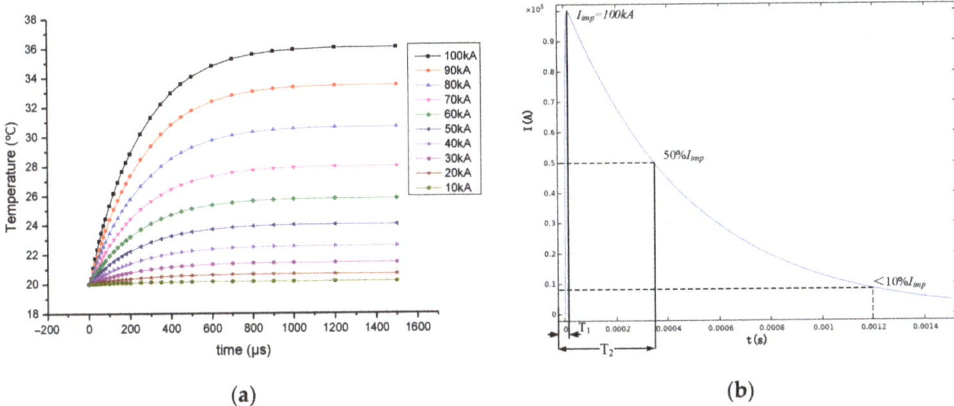

Figure 10. Temperature rise of BIPV at different time and current amplitude. (**a**) The temperature rise of BIPV (**b**) The lightning impulse current waveform.

3.2. Experimental Study

3.2.1. Experimental Method

After multi-physical field simulation analysis, Sample C was selected for experimental verification. As shown in Figure 11, a large current experimental platform for simulating lightning stroke was set up in the laboratory. Sample C was horizontally placed on an insulating table. One copper wire was used to connect the output end of Impulse Current Generator (ICG) to the expected lightning striking point on the metal frame. Another copper wire was used to connect the other end of the metal frame to the grounding electrode.

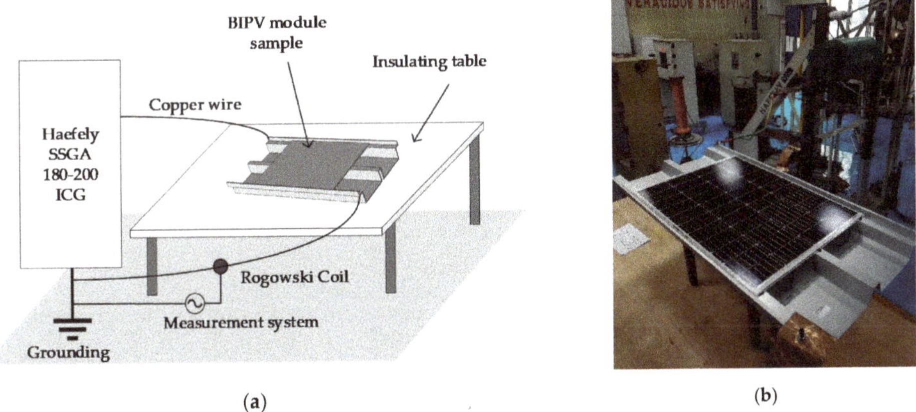

Figure 11. Lightning current withstand performance experiment (**a**) Schematic diagram of experimental platform; (**b**) Layout of experimental setup.

According to the IEC standard [21], the lightning conductor of the highest grade (H-grade) should withstand the impulse current with 100 kA peak value and 2.5 MJ/Ω specific energy, which represents the equivalent energy of the first return stroke component (FRSC) of nature lightning. Since the metal frame of the BIPV module is easy to be struck by lightning, it can function as the lightning receptor on the buildings. The PV module frame should also meet the H-grade impulse current withstand level. In this paper, Haefely SSGA 180–200 ICG was used to output the impulse current with 100 kA peak value and

2.5 MJ/Ω specific energy. Each sample receives three shots. During each shot, the current waveform was measured by a Rogowski coil. The status of the metal frame is observed after each shot to check whether it is damaged.

3.2.2. Experimental Results and Analysis

According to the experimental method of Section 3.2.1, a large impulse current experimental platform was built. The impulse current with amplitude of nearly 100 kA and waveform of 10/350 µs was applied to Sample C by using Haefely SSGA 180–200 ICG. The first impulse current waveform is shown in Figure 12. Table 2 illustrates the parameters of the lightning current in three shots. The waveform of lightning current is defined as T_1/T_2, where T_1 indicates the peak time of lightning current and T_2 indicates the time from the beginning to the attenuation to 1/2 amplitude. Transfer charge Qs is the time integral of lightning current in duration; Specific energy W/R is the time integral of lightning current squared in duration.

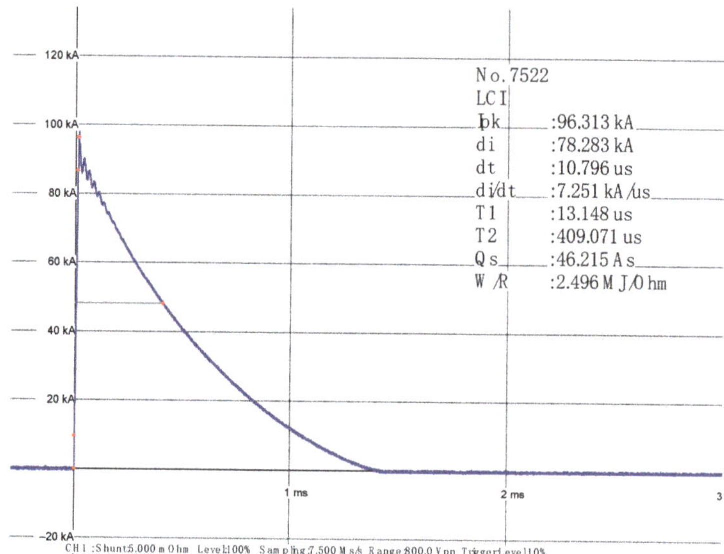

Figure 12. The first impulse current waveform.

Table 2. Impulse current parameter.

No.	Peak Current/kA	T_1/T_2 (µs)	Transfer Charge/As	Specific Energy/(MJ * Ω^{-1})
1	96.313	13.1/409.1	46.251	2496
2	96.508	12.9/397.1	44.674	2426
3	96.938	13.1/400.4	45.073	2451

Due to the impedance of the experimental loop, the impulse currents flowing through the metal frame were a little smaller than 100 kA, i.e., 96.313 kA, 96.508 kA and 96.938 kA respectively. The specific energies of the impulse currents flowing through the metal frame were 2496 MJ/Ω, 2426 MJ/Ω and 2451 MJ/Ω respectively. After three shots, the experimental results showed that the metal frame of Sample C had no signs of deformation and solar cells were not damaged, which was consistent with the results of the thermoelectric coupling analysis model. This shows that the metal frame of BIPV module has adequate lightning energy withstand capability.

4. Conclusions

This paper mainly analyzes the lightning attachment characteristics and lightning energy withstand capability of BIPV modules by numerical simulation and experiment. Two conclusions are drawn as follows,

(a) The electrostatic field model when lightning downward leader above BIPV was built and analyzed with FEM software. An experiment was conducted to verify the numerical analysis. The numerical and experimental study of lightning attachment characteristics shows that the upper edge of the metal frame of BIPV modules is easier to gather charge which can enhance the electric field strength. The electric field strength on the surface of the BIPV module is 132% higher than that of framed double-glass PV module. Therefore, the BIPV module more easily intercepts lightning and the protection efficiency is improved by 114%.

(b) The thermoelectric coupling analysis model is established to solve the potential, current and temperature of lightning BIPV modules during the process of lightning current flows through the metal frame of the BIPV module. The results show that the metal frame of the BIPV modules will not be damaged by the temperature rise of 16.07 °C. In the large impulse current experiment, the metal frame of the BIPV module meets the H-level (the highest level) requirements of the lightning conductor. The BIPV module is not damaged in the experiment which is consistent with the numerical simulation results.

In a word, the metal frame of BIPV can not only intercept lightning stroke to protect the solar cells inside effectively, but can also function as a lightning receptor and down conductor of a building roof. With its good lightning protection performance, BIPV has good application prospects in the rapidly developing distributed PV stations. It has prodigious potential for building integration, especially for less-energy hungry, zero-energy, sustainable, green, and aesthetic building integration. In the future work, our study will focus on the BIPV array on roofs and in facades, evaluating the risk of lightning stroke to BIPV in different application modes. The following research will provide lightning protection methods for green buildings and PV systems.

Author Contributions: Conceptualization, Q.Z. and X.B.; methodology, Q.Z. and X.B.; software, Y.Z. and T.C.; validation, Q.Z. and B.W.; formal analysis, Q.Z. and B.W.; investigation, Y.Z. and T.C.; resources, Q.Z. and B.W.; data curation, X.B.; writing—original draft preparation, Y.Z. and T.C.; writing—review and editing, Q.Z. and Y.Z.; visualization, Y.Z.; supervision, X.B.; project administration, Q.Z. and B.W.; funding acquisition, Q.Z. All authors have read and agreed to the published version of the manuscript.

Funding: This work was supported in part by the Shanghai Municipal Market Supervision and Administration Bureau under Grants 19TBT018 and 20TBT010, and in part by the Shanghai Science and Technology Commission Project under grant 19020500800.

Institutional Review Board Statement: Not applicable.

Informed Consent Statement: Not applicable.

Data Availability Statement: Not applicable.

Conflicts of Interest: The authors declare no conflict of interest.

References

1. Ghosh, A. Potential of Building Integrated and Attached/Applied Photovoltaic (BIPV/BAPV) for Adaptive Less Energy-Hungry Building's Skin: A Comprehensive Review. *J. Clean. Prod.* **2020**, *276*, 123343. [CrossRef]
2. Chul-Sung, L.; Hyo-mun, L.; Min-joo, C.; Yoon, J. Performance Evaluation and Prediction of BIPV Systems under Partial Shading Conditions Using Normalized Efficiency. *Energies* **2019**, *12*, 273527.
3. Barone, G.; Buonomano, A.; Cesare, F.; Forzano, C.; Giuzio, G.F.; Palombo, A. Passive and active performance assessment of building integrated hybrid solar photovoltaic/thermal collector prototypes: Energy, comfort, and economic analyses. *Energy* **2020**, *209*, 118435. [CrossRef]

4. Escarre, J.; Li, H.-Y.; Sansonnens, L.; Galliano, F.; Cattaneo, G.; Heinstein, P.; Nicolay, S.; Bailat, J.; Eberhard, S.; Ballif, C.; et al. When PV modules are becoming real building elements: White solar module, a revolution for BIPV. In Proceedings of the 2015 IEEE 42nd Photovoltaic Specialist Conference (PVSC), New Orleans, LA, USA, 14–19 June 2015; pp. 1–2. [CrossRef]

5. López, C.S.P.; Bonomo, P.; Frontini, F.; Medici, V.; Nespoli, L. Performance assessment of a BIPV Roofing Tile in outdoor testing. In Proceedings of the 2017 IEEE 44th Photovoltaic Specialist Conference (PVSC), Washington, DC, USA, 25–30 June 2017; pp. 2118–2123. [CrossRef]

6. AlRashidi, H.; Issa, W.; Sellami, N.; Ghosh, A.; Mallick, T.K.; Sundaram, S. Performance Assessment of Cadmium Telluride-Based Semi-Transparent Glazing for Power Saving in Façade Buildings. *Energy Build.* **2020**, *215*, 109585. [CrossRef]

7. Yu, S.; Xu, W.; Shi, Z. Study on Application of Random Bidirectional Discharge Model in Risk Assessment of Lightning Disaster at Airport. In Proceedings of the 2018 34th International Conference on Lightning Protection (ICLP), Rzeszow, Poland, 2–7 September 2018; pp. 1–4. [CrossRef]

8. Jiang, T. Electrical Properties Degradation of Photovoltaic Modules Caused by Lightning Induced Voltage. Ph.D. Thesis, Mississippi State University, Starkville, MS, USA, May 2014.

9. Guo, F.-Y.; Wang, Y.; Huang, M.-D. Design of Photovoltaic Building Lightning Protection Equipment Monitoring and Early Warning System. In Proceedings of the 26th Chinese Control and Decision Conference (2014 CCDC), Changsha, China, 31 May–2 June 2014; pp. 2902–2906.

10. Fallah, N.; Gomes, C.; Ab Kadir, M.Z.A.; Nourirad, G.; Baojahmadi, M.; Ahmed, R.J. Lightning protection techniques for roof-top PV systems. In Proceedings of the 2013 IEEE 7th International Power Engineering and Optimization Conference (PEOCO), Langkawi, Malaysia, 3–4 June 2013; pp. 417–421. [CrossRef]

11. Benesova, Z.; Haller, R.; Birkl, J.; Zahlmann, P. Overvoltages in photovoltaic systems induced by lightning strikes. In Proceedings of the 2012 International Conference on Lightning Protection (ICLP), Vienna, Austria, 2–7 September 2012; pp. 1–6. [CrossRef]

12. Phanthuna, N.; Thongchompoo, N.; Plangklang, B.; Bhumkittipich, K. Model and experiment for study and analysis of Photovoltaic lightning effects. In Proceedings of the 2010 International Conference on Power System Technology, Hangzhou, China, 24–28 October 2010; pp. 1–5. [CrossRef]

13. Karim, M.R.; Ahmed, M.R. Lightning Effect on a Large-Scale Solar Power Plant with Protection System. In Proceedings of the 2019 1st International Conference on Advances in Science, Engineering and Robotics Technology (ICASERT), Dhaka, Bangladesh, 3–5 May 2019; pp. 1–5. [CrossRef]

14. Jia, Y.G.; Liu, H.Y.; Su, S.F.; Ding, S.L.; Han, Y.Q.; Liu, Y.; Zhang, X.S. Integrated Photovoltaic System Design and Key Points Analysis Based on Green Building. *Appl. Mech. Mater.* **2014**, *641*, 994–998. [CrossRef]

15. Guo, F.Y.; Wang, Y.; Huang, M.D.; Yue, W. Fault Tree Establishment of Lightning Protection System Safety of Solar Photovoltaic Building. *Adv. Mater. Res.* **2013**, *860*, 210–213. [CrossRef]

16. Zaini, N.H.; Ab-Kadir, M.Z.A.; Izadi, M.; Ahmad, N.I.; Radzi, M.A.M.; Azis, N.; Hasan, W.Z.W. On the effect of lightning on a solar photovoltaic system. In Proceedings of the 2016 33rd International Conference on Lightning Protection (ICLP), Estoril, Portugal, 25–30 September 2016; pp. 1–4. [CrossRef]

17. Cho, J.; Kim, J.; Lee, T.-K.; Kim, K.-H.; Woo, J.-W. Analysis of Lightning Overvoltage According to Position of Lightning-Induced Voltage at the Solar Power Plant. In Proceedings of the 2019 11th Asia-Pacific International Conference on Lightning (APL), Hong Kong, 12–14 June 2019; pp. 1–4. [CrossRef]

18. Dechthummarong, C.; Thepa, S.; Chenvidhya, D.; Jivacate, C.; Kirtikara, K.; Thongpron, J. Lightning impulse test of field-aged PV modules and simulation partial discharge within MATLAB. In Proceedings of the 2012 9th International Conference on Electrical Engineering/Electronics, Computer, Telecommunications and Information Technology, Phetchaburi, Thailand, 16–18 May 2012; pp. 1–4.

19. Tamura, K.; Araki, K.; Kumagai, I.; Nagai, H. Lightning test for concentrator photovoltaic system. In Proceedings of the 2011 37th IEEE Photovoltaic Specialists Conference, Seattle, WA, USA, 10–15 June 2011; pp. 000996–000998. [CrossRef]

20. Coetzer, K.M.; Wiid, P.G.; Rix, A.J. Investigating Lightning Induced Currents in Photovoltaic Modules. In Proceedings of the 2019 International Symposium on Electromagnetic Compatibility—EMC EUROPE, Barcelona, Spain, 2–6 September 2019; pp. 261–266. [CrossRef]

21. IEC 62561-2:2018. Lightning Protection System Components (LPSC)—Part 2: Requirements for Conductors and Earth Electrodes. Available online: https://webstore.iec.ch/publication/29398 (accessed on 25 January 2018).

22. Golde, R.H.; Perry, F.R. The frequency of occurrence and the distribution of lightning flashes to transmission lines. *Trans. Am. Inst. Electr. Eng.* **1945**, *64*, 902–910. [CrossRef]

23. Young, F.S.; Clayton, J.M.; Hileman, A.R. Shielding of transmission lines. *IEEE Trans. Power Appar. Syst.* **1963**, *82*, 132–154.

24. Love, E.R. Improvements in Lightning Stroke Modeling and Applications to Design of EHV and UHV Transmission Lines. Master's Thesis, University of Colorado, Boulder, CO, USA, 1973.

25. Whitehead, E.R. Cigre survey of the lightning performance of extra-high-voltage transmission lines. *Electra* **1974**, *33*, 63–89.

26. Anderson, J.D. *Transmission Line Reference Book-345 kV and above*; Electric Power Research Institute: Palo Alto, CA, USA, 1982; pp. 142–144.

27. IEEE Std 998-2012, IEEE Guide for Direct Lightning Stroke Shielding of Substations. Available online: https://ieeexplore.ieee.org/document/6656813 (accessed on 30 April 2013).

28. Gamerota, W.R.; Elismé, J.O.; Uman, M.A.; Rakov, V.A. Current waveforms for lightning simulation. *IEEE Trans. Electromagn. Compat.* **2012**, *54*, 880–888. [CrossRef]
29. Berger, K.; Anderson, R.B.; Kroninge, R.H. Parameters of Lighting Flashes. *Electra* **1975**, *41*, 23–37.
30. Bruce, C.E.R.; Golde, R.H. The lightning discharge. *Inst. Elect. Eng.* **1941**, *88*, 487–505.

Article

A Zero Crossing Hybrid Bidirectional DC Circuit Breaker for HVDC Transmission Systems

Geon Kim, Jin Sung Lee, Jin Hyo Park, Hyun Duck Choi * and Myoung Jin Lee *

Department of ICT Convergence System Engineering, Chonnam National University, Gwangju 61186, Korea;
rjsdlchd@naver.com (G.K.); lsc5176@naver.com (J.S.L.); wlsgy6691@naver.com (J.H.P.)
* Correspondence: ducky.choi@jnu.ac.kr (H.D.C.); mjlee@jnu.ac.kr (M.J.L.); Tel.: +82-62-530-1810 (M.J.L.)

Abstract: With the increasing demand for renewable energy power generation systems, high-power DC transmission technology is drawing considerable attention. As a result, stability issues associated with high power DC transmission have been highlighted. One of these problems is the fault current that appears when a fault occurs in the transmission line. If the fault current flows in the transmission line, it has a serious adverse effect on the rectifier stage, inverter stage and transmission line load. This makes the transmission technology less reliable and can lead to secondary problems such as fire. Therefore, fault current must be managed safely. DC circuit breaker technology has been proposed to solve this problem. However, conventional technologies generally do not take into account the effects of fault current on the transmission line, and their efficiency is relatively low. The purpose of this study is to introduce an improved DC circuit breaker that uses a blocking inductor to minimize the effect of fault current on the transmission line. It also uses a ground inductor to efficiently manage the LC resonant current and dissipate residual current. DC circuit breakers minimize adverse effects on external elements and transmission lines because the use of elements placed on each is distinct. All of these processes are precisely verified by conducting simulation under 200 MVA ($\pm$100 kV) conditions based on the VSC-based HVDC transmission link. In addition, the mechanism was explained by analyzing the simulation results to increase the reliability of the circuit in this paper.

Keywords: DC circuit breaker; $\pm$230 kV MMC-HVDC; zero-crossing DCCB; DC transmission line; fault current; hybrid DCCB; bidirectional DCCB; external elements; energy dissipation

Citation: Kim, G.; Lee, J.S.; Park, J.H.; Choi, H.D.; Lee, M.J. A Zero Crossing Hybrid Bidirectional DC Circuit Breaker for HVDC Transmission Systems. *Energies* **2021**, *14*, 1349. https://doi.org/10.3390/en14051349

Academic Editor: Ahmed Abu-Siada

Received: 30 December 2020
Accepted: 25 February 2021
Published: 2 March 2021

1. Introduction

Modern society needs much more electric power than in the past [1]. Therefore, the power generation sector is focused on power generation and transmission [2]. As a result, the instability of high-power direct current (DC) transmission technology was emphasized [3].

High voltage direct current (HVDC) transmission systems can be divided into voltage source converter (VSC) and current source converter (CSC) HVDC systems according to the switching element of the converter [4]. VSC HVDCs using insulated gate bipolar transistor (IGBT) have a small installation area and does not require reactive power compensation facilities [5]. It can also be used in a variety of ways because it has free bidirectional transmission and does not require an alternating current (AC) voltage source [6].

If a problem occurs in the transmission line of the VSC HVDC, a fault current is generated by the fast waveform [7]. This can cause the entire grid of multi-terminal HVDC to be blocked or severely adversely affect the power elements connected to the transmission line [8,9]. Therefore, DC circuit breaker technology that can safely manage fault current has been proposed [10].

DC circuit breaker technology can be divided several ways, of which the zero-crossing breaking method safely manages fault current by generating zero current in the main transmission line and turning off the switch [11,12]. The performance of a DC circuit breaker is an indicator of the breaking time of the main transmission line and the amount

185

of leakage current to the outside [13–15]. Therefore, it is very important to manage electric current on the transmission line elements during high-power transmission [16].

DC circuit breakers (DCCBs) are generally classified by switch type into mechanical DCCB, semiconductor DCCB, and hybrid DCCB [17]. Hybrid DCCB is considered the most suitable type of DC circuit breakers for HVDC because it solves the problem of slow operation speed, the disadvantage of mechanical DCCB, and power loss, the disadvantage of semiconductor DCCB [18]. In general, hybrid DCCB places a mechanical switch on the main transmission line through steady-state current, so the current does not pass through the semiconductor switch in the on-state [19].

The breaking direction of the DC circuit breaker can be either unidirectional or bidirectional [20]. In a bidirectional DCCB, the breaker can operate normally regardless of the location of the main transmission line, which facilitates mass production and reduces manufacturing costs [10]. In this paper, we simulated a hybrid bidirectional HVDC DCCB using a zero-crossing method assuming VSC-based HVDC transmission.

Section 2 describes the mechanism of current flow through the circuit operation process and how it is used as a bidirectional DCCB. Section 3 describes the simulation results and analysis. Initially, detailed simulation conditions are specified, and the reverse charging process, blocking inductor and ground inductor are analyzed in order from Sections 3.1–3.3. In Section 3.4 we will analyze the use of switching elements and energy dissipation to increase the reliability of the DCCB and conclude in Section 4.

2. Circuit Operation Process

Figure 1 shows the current flow in the DCCB proposed in this paper is in a steady-state. If the DC steady-state persists, the blocking inductor has no effect. It does not cause any additional power loss. The steady-state current charges the reverse charge capacitor, so no further action is required when operating later. Even in this case, there is no power loss when the reverse charge thyristor is fully charged. The proposed DCCB works well even if there is a fault current anywhere in the transmission line.

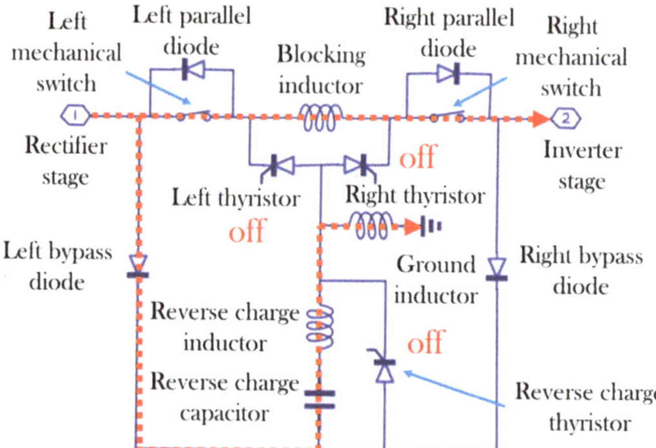

Figure 1. The overall circuit diagram of the DCCB proposed in this paper and the steady-state current flow.

Figure 2 shows the flow of the fault current from the rectifier stage to the DCCB and the emitted the inductor and the capacitor (LC) resonant current. When a fault current is detected, it sends a signal to the gate terminal of the reverse charge thyristor and proceeds with the reverse charging process. At this time, some of the fault currents are used in the reverse charging process to create a larger resonant current. Immediately after that, the

LC resonant current is emitted by turning of the left or right thyristors according to the direction of the fault current [21]. When the fault current and LC resonant current are zero-crossed, the mechanical switch on the main transmission line turns off to eliminate the effect of fault current.

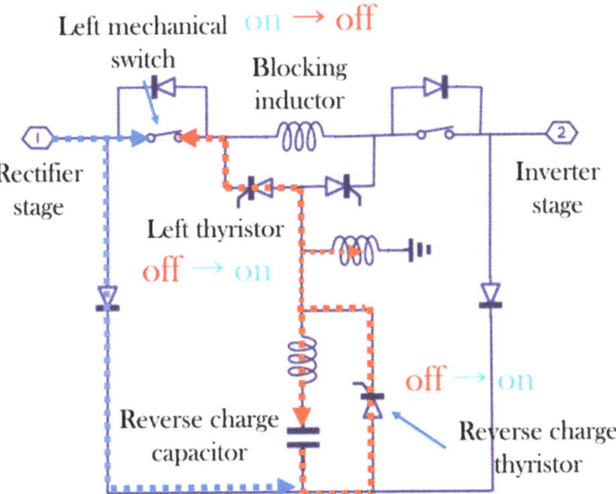

Figure 2. Current flow of LC resonant emission current and fault current from the rectifier stage to the DCCB.

The DCCB in this paper can break in both directions, as described above. Unlike the example in Figure 2 where there is a fault current in the rectifier stage, when the fault current flows from the inverter stage to the DCCB, the reverse charging process is the same, and when the right thyristor is turned on, the LC resonant current is emitted to generate a zero crossing point and the right mechanical switch is turned off.

Figure 3 shows the flow of residual current after zero-crossing of the DCCB main switch. To minimize the effect of residual current on the rectifier and inverter stages, the current is returned to the DCCB through parallel diodes and bypass diodes. At this time, the reverse charge thyristor is turned on continuously, dissipating the energy in the DCCB to the ground. When the residual current is removed to the ground, it resonates through the ground inductor passing through, leaving a small amount of energy in the DCCB but with little effect.

Figure 3a shows the current flow after breaking the fault current passing from the rectifier stage to the DCCB, and Figure 3b shows the current flow after breaking the fault current passing from the inverter stage to the DCCB. The mechanism for directing current flow through a diode connected in parallel to the mechanical switch, which is usually the main switch, is the same. Depending on the direction of the fault current according to the placement of the DCCB, the flow of residual current is symmetrical and the process of dissipating energy through the ground is the same.

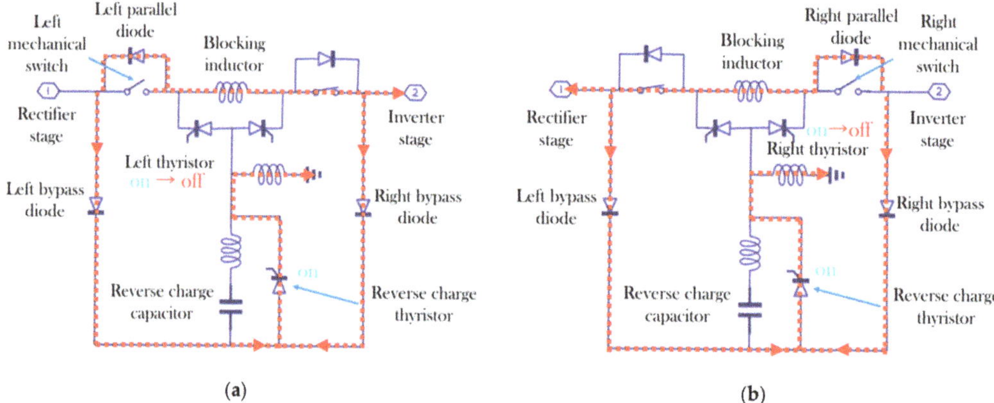

Figure 3. Flow of residual current after zero-crossing of the main switch of the proposed DCCB: (**a**) Current flow after breaking the fault current passing from the rectifier stage to the DCCB; (**b**) Current flow after breaking the fault current passing from the inverter stage to the DCCB.

3. Simulation Results and Analysis

Figure 4a schematically shows the simulated VSC-based transmission circuit to test the DCCB proposed in this paper. The simulation was performed assuming that a fault occurs in 230 kV, 50 Hz, and 200 MVA transmission between the DCCB and a 30 km cable [22,23]. Assuming the fault current flowing to the DCCB in the rectifier stage in the schematic, the DCCB in this paper can break the fault current in both directions, so in the example, the DCCB could be placed on the adjacent stage of the modular multilevel converter (MMC) on the right.

Figure 4b is a graph showing the magnitude of the fault current when a fault occurs within 3 s under a given simulation condition. Simulations performed under these conditions show that the current increases rapidly in 3 s and then gradually decreases [24].

The basic parameters used in the simulation are summarized in Tables 1–3. These parameters are default values. Optimized values are identified by analyzing the results as described later. Also, the current limiting reactors in Table 2, not shown in the schematic diagram of Figure 2, refer to the inductors used at both ends of the transmission line of a general DCCB.

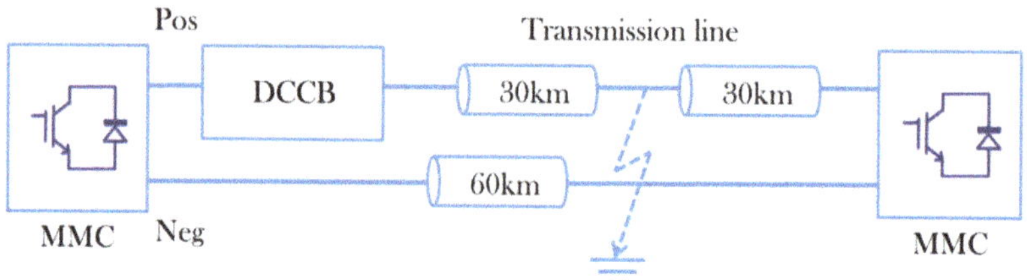

Figure 4. *Cont.*

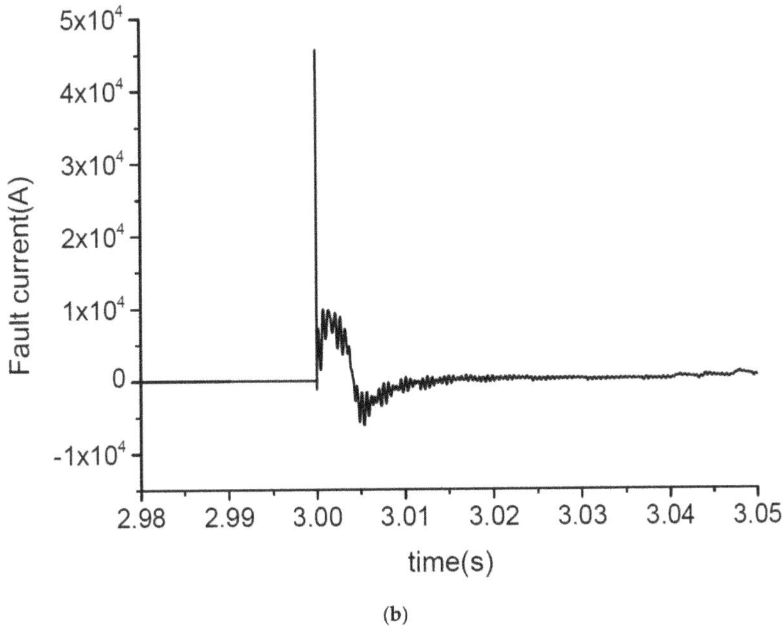

(b)

Figure 4. DCCB simulation condition proposed in this paper (VSC-Based HVDC transmission line) (**a**) Schematic diagram the DCCB simulation; (**b**) Magnitude of the fault current (Fault current occurrence time = 3.00 s).

Table 1. Transmission line parameters used to simulate the proposed DCCB.

Data	Length [km]	R [Ω/km]	L [mH/km]	C [μF/km]
Positive pole cable	30	0.0139	0.159	0.231
Negative pole cable	60	0.0139	0.159	0.231

Table 2. Basic characteristics of the elements used to simulate the proposed DCCB.

Data	Parameter
Blocking inductor	1 mH
Reverse charge inductor	1 mH
Reverse charge capacitor	1 mF
Ground inductor	1 mH
Current limiting reactor (Here, the inductor installed at both ends of the DCCB.)	1 mH

Table 3. Characteristics of the switching elements used to simulate the proposed DCCB.

Data	Internal Resistance Ron [mΩ]	Snubber Resistance Rs [kΩ]	Snubber Capacitance Cs [μF]
Mechanical switch (ideal)	1	100	Inf
Diode	1	0.5	0.25
Thyristor	1	0.5	0.25

3.1. Effectiveness of the Reverse Charging Process

Figure 5 is a graph comparing the reverse-charge method and the normal-charge method used in the proposed DCCB. The resonant current signals the switch's gate (in this case, the left thyristor), assuming that it is emitted within 0 s when turned on. In general, the larger the value of the resonant current emitted, the faster the same amount of fault current can be zero-crossed. However, as the amount of generated resonant current increases, more energy remains, making it difficult to eliminate residual current.

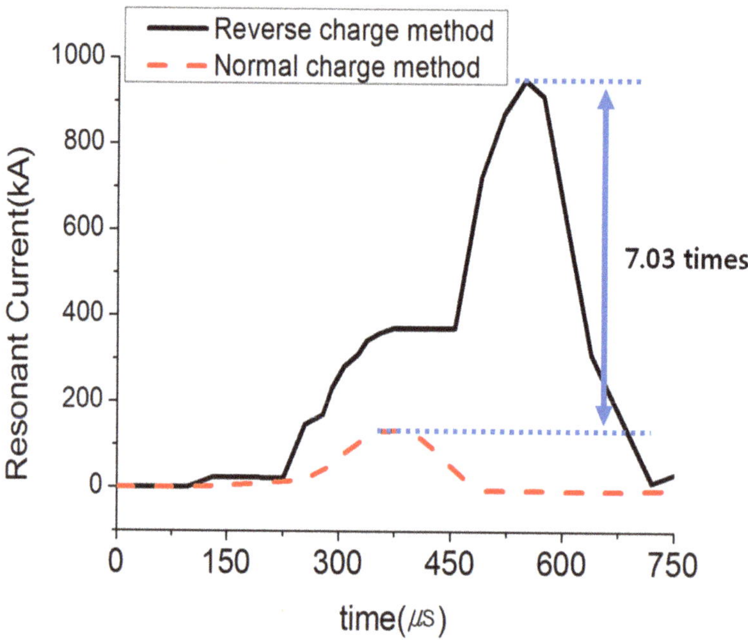

Figure 5. Comparison of the magnitude of resonant current between the reverse-charge method and the normal-charge method in the DCCB in the proposed DCCB.

When comparing the reverse-charge method and the normal-charge method used in the proposed DCCB, the reverse-charge method generates a peak resonant current that is 7.03 times larger than that of the normal-charge method. The slope of the current before peaking is also steeper when using the reverse-charge method than the normal-charge method [25]. This means that when using the reverse-charge method, zero-crossing can be performed faster on the mechanical switch of the main current branch. However, a corresponding problem arises, more energy must be dissipated. This issue will be addressed in Section 3.4.2, but in the case of the DCCB in this paper, energy can be dissipated efficiently due to the path of current to ground.

3.2. Blocking Inductor

Figure 6 shows the peak current in the inverter and rectifier stages depending on the blocking inductor. Regardless of the value of the two DCCB inductors, the impedance of the transmission line increases as the value of the blocking inductor increases. Therefore, the amount of current flow in the inverter stage is reduced. The graph does not clearly show the effect of the inductor when the blocking inductor is less than 10^{-4} H. However, when it exceeds 10^{-3} H, the peak current is greatly improved. When the value of the blocking inductor exceeds 10^{-3} H, the peak current in the inverter stage decreases rapidly.

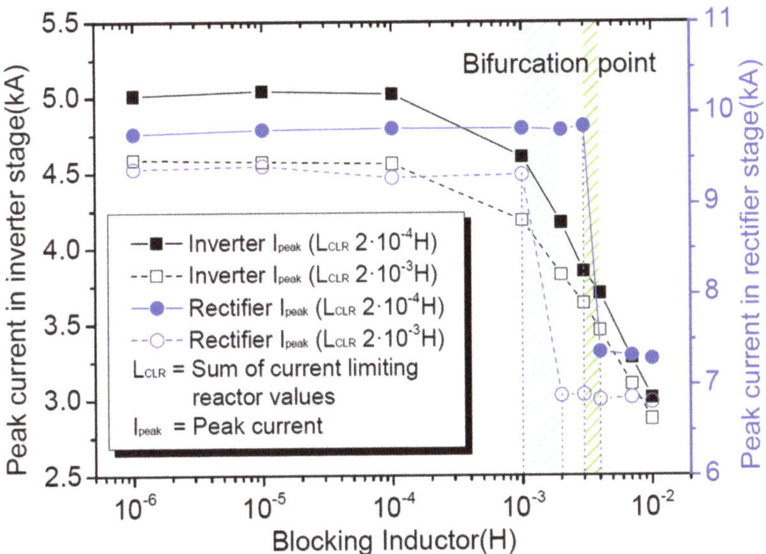

Figure 6. Graph of the peak current in the inverter and rectifier stages depending on the blocking inductor.

The peak current in the rectifier stage decreases significantly after passing the bifurcation point. This bifurcation point occurs at 4 mH (4×10^{-3} H) when the sum of current limiting reactor values is 0.2 mH (2×10^{-4} H), and at 2 mH (2×10^{-3} H) when the sum of current limiting reactor values is 2 mH (2×10^{-3} H). This kind of bifurcation point occurs when the value of the blocking inductor is greater than the value of current limiting reactors and the sum of the inductors of the main current branch exceeds 3 mH (3×10^{-3} H). Increasing the value of the blocking inductor and exceeding the aforementioned bifurcation point can rapidly reduce the peak current in the rectifier stage.

However, since the blocking inductor is located in the transmission line, if an excessively large value is used, the time from initial transmission to steady-state may be extended. Therefore, the optimized value of the blocking inductors must cross the bifurcation point to rapidly reduce the peak current in the rectifier stage so as not to be excessive.

3.3. Ground Inductor

Figure 7 shows the peak current, zero crossing time (ZCT), and rectifier current stable time (RST) depending on the ground inductor. ZCT is the time the main switch is turned off at the main current branch. The fault current is equal to the LC resonant current, so this generates a current zero. RST is the time until the current in the rectifier stage is stabilized (usually less than the steady-state current).

Increasing the ground inductor causes less LC resonant current to flow to ground and more LC resonant current to emit to the main switch. Therefore, ZCT is shortened because more resonant current is emitted. However, the larger ground inductor, the less the current returned to the DCCB by increasing its own impedance after the main switch breaks. As a result, the ground current decreases and the rectifier stage takes a long time to stabilize.

In the graph, when the value of the ground inductor is small, the characteristic change in characteristics is very small, but when it is greater than 10^{-3} H, the effect is remarkable, so the value of the ground inductor has a trade-off relationship between ZCT and RST. Since ZCT and RST cannot be designed well at the same time, the trade-off characteristics must be considered in the DCCB design procedure.

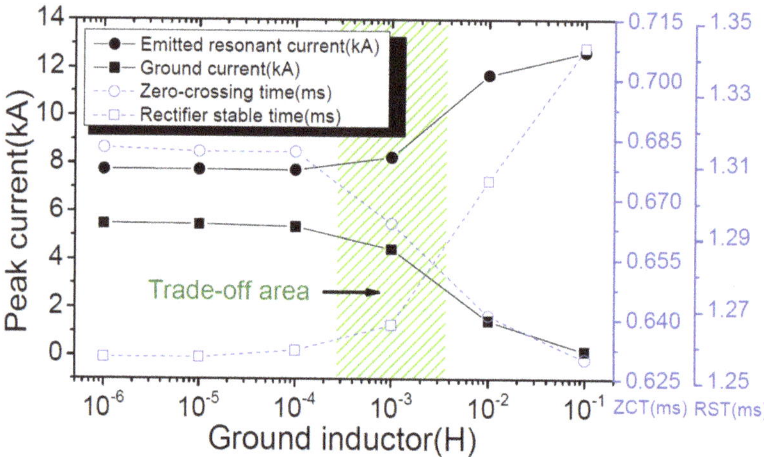

Figure 7. The emitted resonant current, ground current, zero-crossing time (ZCT), and rectifier stable time (RST) depending on the ground inductor.

The ground inductor has the same characteristics when changed to a resistive element. Even in the case of the resistors, there is no significant power loss when the reverse charge capacitor is fully charged. Likewise, the larger the resistance, the more LC resonant current is emitted to the main switch, but less current is returned to the DCCB after the main switch break. However, when using an inductor on the ground, the more the current changes over time, the higher the impedance and the less current flow. Therefore, less current flows to the ground than when using a resistor, and conversely, more resonant current can be effectively emitted to the main switch. Because of these characteristics, in this paper, we use a ground inductor for the DCCB with better characteristics.

3.4. Elements

3.4.1. Switching Element Verification

Figure 8 shows the peak current in the inverter and rectifier stages when different cases. In case 1, all elements are in normal conditions in the DCCB. In case 2, the left and right parallel diodes are removed. In case 3, the reverse charging process, which is one of the use of the left and right bypass diodes, proceeds the same as in case 1, except that residual current flows in the DCCB after the main switch breaks. Case 4 assumes that both the left and right thyristors are both turned on and the LC resonant current is emitted without specifying the direction.

In case 2, the load on the parallel diodes disappears. Therefore, the influence of the residual current increases in the rectifier stage. In the DCCB in this paper, the peak current in the rectifier stage can be reduced by about 28% using parallel diodes. This can significantly reduce the adverse effects on the elements of the rectifier stage. There doesn't seem to be a significant difference in the result value for case 3 compared to case 1, but there is a problem with energy dissipation. This is described in Section 3.4.2. By using of the bypass diode, the residual energy caused by the LC resonant current can be efficiently dissipated. In case 4, the direction of the LC resonant current cannot be specified, since both thyristors are turned on. Therefore, part of the LC resonant current affects the inverter stage and creates a high load on the transmission line. In this paper, specifying the emission direction of the LC resonant current in the DCCB can reduce peak current in the inverter stage by about 17%. This reduction in current significantly reduces the load on the transmission line, mainly in the inverter stage.

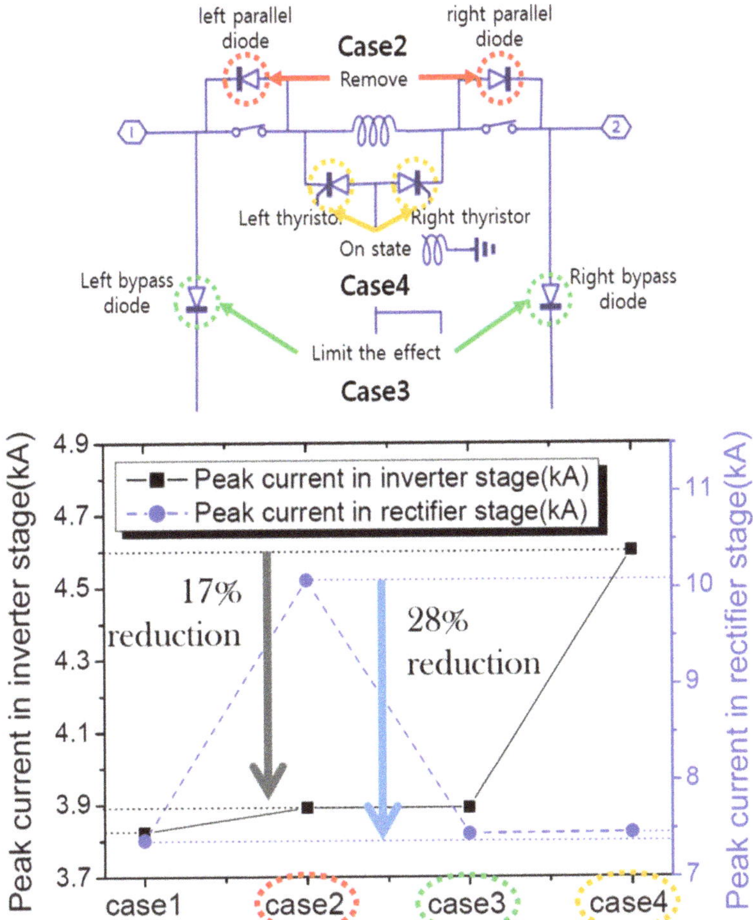

Figure 8. Peak current in the inverter and rectifier stage when different cases; Case 1: All elements are operating under normal conditions; Case 2: Under normal conditions, the parallel diodes are removed; Case 3: Under normal condition, the bypass diode is changed to a thyristor, the mechanical switch is turned off, the changed thyristor is also turned off (The reverse charging process works the same.); Case 4: Under normal conditions, resonant current is emitted without specifying the direction of the thyristor (both the left and right thyristors are turned on).

3.4.2. Energy Dissipation Verification

Figure 9a schematically shows the dissipation process in the proposed DCCB. The use of dissipative thyristors near ground in this DCCB is schematically illustrated in Figure 9c,d. Also, the measured node was expressed to verify the energy dissipation between the DCCB and the transmission line. The voltage measurement in Figure 9c represents the residual energy of the DCCB, and the residual energy of the transmission line is verified by the voltage measurement in Figure 9d.

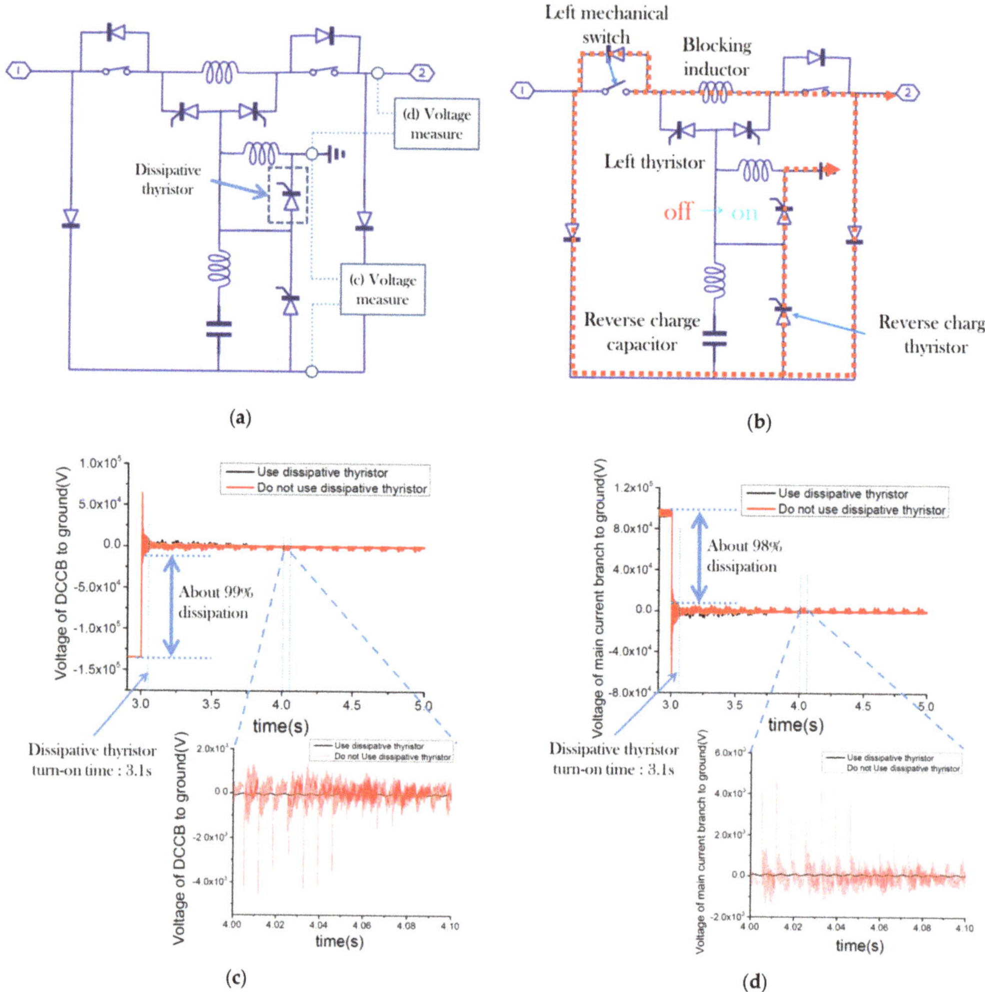

(a) **(b)** **(c)** **(d)**

Figure 9. Simulation of energy dissipation in the DCCB proposed in this paper (**a**) Schematic illustration of the DCCB diagram in the presence of the dissipative thyristor. (**b**) Current flow in the presence of the dissipative thyristor. (**c**) Graph of voltage between the DCCB and ground over time. (**d**) Graph of voltage between the main current branch and ground over time.

Figure 9b schematically shows the current flow after the main mechanical switch is turned off when there is a dissipative thyristor in Figure 9a. In the absence of a dissipative thyristor in the proposed DCCB, residual current will inevitably flow through the ground inductor to ground, resulting in ripple. As can be seen in Figure 9c,d, the value of this ripple is very small compared to the steady-state voltage. If you need to further reduce the ripple, turn on the dissipative thyristor after the main switch is turned off, so the current will not pass through the ground inductor. As shown in Figure 3b, when the fault current occurs in the opposite direction, as shown in Figure 3b, only the current flow from the DCCB to the ground passes through the dissipative thyristor.

Figure 9c shows the voltage between the DCCB and the ground. In the proposed DCCB, ZCT occurs within 1ms under simulated conditions and RST is less than 2 ms. Therefore, in order to independently analyze the use of the dissipative thyristor in the energy dissipation process, the dissipative thyristor was turned on after 0.1 s (100 ms). Since the fault current did not occur 3 s ago, the voltage of the DCCB is the same as the voltage charged to the reverse charge capacitor under the normal steady-state. When the main switch is turned off and the energy dissipation process is reached, 99% of the voltage is lost to ground even without the dissipative thyristor, leaving little residual energy in the DCCB. Therefore, even without the surge arrester, the energy dissipation problem rarely occurs when the DCCB proposed in this paper is used. This small ripple can also be made to converge to zero by turning on the dissipative thyristor shown in Figure 9a. Analyzing the current at 4–4.1 s confirms that it converges to zero when the dissipative thyristor is turned on.

Figure 9d shows the voltage between the main current branch and ground. Figure 9d is very similar to the graph in Figure 9c. The voltage state is normal 3 s ago, when the main switch is turned off, the voltage decreases sharply. At this time, the proposed DCCB rapidly dissipates the energy equivalent to 98% of the energy seen under normal voltage conditions, similar to Figure 9c. This amount of loss doesn't cause much trouble but using a dissipative thyristor to converge the voltage to zero can completely eliminate the energy dissipation problem.

4. Conclusions

We have designed an improved DC circuit breaker taking into account external elements and the transmission line by expanding the conventional breaker technology. Due to the development of high-power transmission, the influence of the current on the external elements and the transmission line cannot be ignored. The DC circuit breaker in this paper uses a blocking inductor to reduce the load on the transmission line at the inverter stage and significantly reduce the amount of current flow through the rectifier stage based on the bifurcation point to protect the device and circuit. When determining the value of the ground inductor, ZCT and RST were found to have an inverse relationship, and the inductor was selected according to the optimal analysis methodology for the proposed DCCB design. In addition, verification of the individual switching elements was carried out in detail to confirm the use of each element in the DCCB. Moreover, the problem of energy dissipation, which is a concern due to the lack of a reverse charging process or a surge arrester, was solved by analyzing the ground voltage between the DCCB and the transmission line. The current flow mechanism has been analyzed to increase reliability and aid further research in the field of DCCB.

Author Contributions: Writing—original draft, G.K.; Investigation, J.S.L.; Formal analysis, J.H.P.; Supervision, M.J.L.; Writing—review & editing, H.D.C. All authors have read and agreed to the published version of the manuscript.

Funding: This work was supported in part by the National Research Foundation of Korea (NRF) grant funded by the Korean government (MSIT; grant number 2018R1A2B600821613) and in part by the BK21 FOUR Program(Fostering Outstanding Universities for Research, 5199991714138) funded by the Ministry of Education(MOE, Korea) and National Research Foundation of Korea(NRF).

Institutional Review Board Statement: Not applicable.

Informed Consent Statement: Not applicable.

Data Availability Statement: Not applicable.

Acknowledgments: The EDA tool was supported by the IC Design Education Center(IDEC), Korea.

Conflicts of Interest: The authors declare no conflict of interest.

References

1. Asplund, G. Sustainable energy systems with HVDC transmission. In Proceedings of the Eighth IEEE International Symposium on Spread Spectrum Techniques and Applications—Programme and Book of Abstracts, Sydney, NSW, Australia, 30 August–2 September 2004.
2. Long, W.; Nilsson, S. HVDC transmission: Yesterday and today. *IEEE Power Energy Mag.* **2007**, 5, 22–31. [CrossRef]

3.   Rudervall, R.; Charpentier, J.P.; Raghuveer, D. High voltage direct current (HVDC) transmission systems technology review paper. *Energy Week* **2000**, *2000*, 1–19.

4.   Flourentzou, N.; Agelidis, V.G.; Demetriades, G.D. VSC-Based HVDC power transmission systems: An overview. *IEEE Trans. Power Electron.* **2009**, *24*, 592–602. [CrossRef]

5.   Nakajima, T.; Irokawa, S. A control system for HVDC transmission by voltage sourced converters. In Proceedings of the 199 IEEE Power Engineering Society Summer Meeting. Conference Proceedings (Cat. No.99CH36364), Edmonton, AB, Canada, 18–22 July 1999.

6.   Schettler, F.; Huang, H.; Christl, N. HVDC transmission systems using voltage sourced converters design and applications. In Proceedings of the 2000 Power Engineering Society Summer Meeting (Cat. No.00CH37134), Seattle, WA, USA, 16–20 July 2000.

7.   Elserougi, A.A.; Abdel-Khalik, A.S.; Massoud, A.M.; Ahmed, S. A new protection scheme for HVDC converters against DC-side faults with current suppression capability. *IEEE Trans. Power Deliv.* **2014**, *29*, 1569–1577. [CrossRef]

8.   Mei, J.; Fan, G.; Ge, R.; Wang, B.; Zhu, P.; Yan, L. Research on Coordination and optimal configuration of current limiting devices in HVDC Grids. *IEEE Access* **2019**, *7*, 106727–106739. [CrossRef]

9.   Tzelepis, D.; Blair, S.M.; Dysko, A.; Booth, C. DC busbar protection for HVDC Substations incorporating power restoration control based on dyadic sub-band tree structures. *IEEE Access* **2019**, *7*, 11464–11473. [CrossRef]

10. Franck, C.M. HVDC circuit breakers: A review identifying future research Needs. *IEEE Trans. Power Deliv.* **2011**, *26*, 998–1007. [CrossRef]

11. Shukla, A.; Demetriades, G.D. A survey on hybrid circuit-breaker topologies. *IEEE Trans. Power Deliv.* **2015**, *30*, 627–641. [CrossRef]

12. Van Gelder, P.; Ferreira, J.A. Zero volt switching hybrid DC circuit breakers. In Conference Record of the 2000 IEEE Industry Applications Conference. In Proceedings of the Thirty-Fifth IAS Annual Meeting and World Conference on Industrial Applications of Electrical Energy (Cat. No. 00CH37129), Rome, Italy, 8–12 October 2000.

13. Xu, Z.; Xiao, H.; Xiao, L.; Zhang, Z. DC fault analysis and clearance solutions of MMC-HVDC Systems. *Energies* **2018**, *11*, 941. [CrossRef]

14. Javed, W.; Chen, D.; Farrag, M.E.; Xu, Y. System configuration, fault detection, location, isolation and restoration: A review on LVDC microgrid protections. *Energies* **2019**, *12*, 1001. [CrossRef]

15. Li, B.; He, J.; Li, Y.; Wen, W. A novel DCCB reclosing strategy for the flexible HVDC Grid. *IEEE Trans. Power Deliv.* **2019**, *35*, 244–257. [CrossRef]

16. Meah, K.; Ula, S. Comparative evaluation of HVDC and HVAC transmission systems. In Proceedings of the 2007 IEEE Power Engineering Society General Meeting, Tampa, FL, USA, 24–28 June 2007; pp. 1–5.

17. Wei, T.; Yu, Z.; Zeng, R.; Chen, Z.; Zhang, X.; Wen, W.; Huang, Y. A novel hybrid DC circuit breaker for nodes in multi-terminal DC system. In Proceedings of the IECON 2017—43rd Annual Conference of the IEEE Industrial Electronics Society, Beijing, China, 29 October–1 November 2017.

18. Daibo, A.; Niwa, Y.; Asari, N.; Sakaguchi, W.; Takimoto, K.; Kanaya, K.; Ishiguro, T. High-speed current interruption performance of hybrid DCCB for HVDC transmission system. In Proceedings of the 4th International Conference on Electric Power Equipment—Switching Technology (ICEPE-ST), Xi'an, China, 22–25 October 2017.

19. Peng, C.; Husain, I.; Huang, A.Q.; LeQuesne, B.; Briggs, R. A fast mechanical switch for medium voltage hybrid DC and AC circuit breakers. In Proceedings of the 2015 IEEE Energy Conversion Congress and Exposition (ECCE), Montreal, QC, Canada, 20–24 September 2015.

20. Ray, A.; Rajashekara, K.; Banavath, S.N. Bidirectional coupled inductor based hybrid circuit breaker topologies for DC System Protection. In Proceedings of the 2019 IEEE Applied Power Electronics Conference and Exposition (APEC), Anaheim, CA, USA, 17–21 March 2019; pp. 1138–1145.

21. Sima, W.; Fu, Z.; Yang, M.; Yuan, T.; Sun, P.; Han, X.; Si, Y. A novel active mechanical HVDC Breaker with consecutive interruption capability for fault clearances in MMC-HVDC systems. *IEEE Trans. Ind. Electron.* **2018**, *66*, 6979–6989. [CrossRef]

22. Li, J.; Zhao, X.; Song, Q.; Rao, H.; Xu, S.; Chen, M. Loss calculation method and loss characteristic analysis of MMC based VSC-HVDC system. In Proceedings of the 2013 IEEE International Symposium on Industrial Electronics, Taipei, Taiwan, 29–31 May 2013; pp. 1–6.

23. Amin, M.; Molinas, M.; Lyu, J.; Mohammad, A. Oscillatory phenomena between wind farms and HVDC systems: The impact of control. In Proceedings of the 2015 IEEE 16th Workshop on Control and Modeling for Power Electronics (COMPEL), Vancouver, BC, Canada, 12–15 July 2015.

24. Jia, H.; Yin, J.; Wei, T.; Huo, Q.; Li, J.; Wu, L. Short-circuit fault current calculation method for the multi-terminal DC Grid considering the dc circuit breaker. *Energies* **2020**, *13*, 1347. [CrossRef]

25. Guo, Y.; Wang, G.; Zeng, D.; Li, H.; Chao, H. A thyristor full-bridge-based dc circuit breaker. *IEEE Trans. Power Electron.* **2019**, *35*, 1111–1123. [CrossRef]

MDPI
St. Alban-Anlage 66
4052 Basel
Switzerland
Tel. +41 61 683 77 34
Fax +41 61 302 89 18
www.mdpi.com

Energies Editorial Office
E-mail: energies@mdpi.com
www.mdpi.com/journal/energies